Department o
PSA Specialis

Costs-in-Use Tables

THIRD EDITION 1991

LONDON: HMSO

First published 1977
Third Edition 1991

ISBN 0 11 671531 6

The price of this publication has been set by PSA to reflect the cost of preparation.

USER NOTES:

Contents

Explanation

Introduction

1 The Tables provide data in respect of:
Capital cost
Cleaning cost
Building cyclical maintenance and renewal cost
The cost of repairs to remedy defects in building design, construction or materials
The mass of components
The heat resistance of components

2 This handbook has been prepared to assist designers and cost consultants to take account of Costs-in-Use when selecting materials and components, and when advising clients on lifetime costs.

3 Maintenance costs increase with the age of the building. The type of building also has an effect. The costs indicated in these Tables are the average of all types of building and the average over their lifespan.

4 The left hand page provides costs-in-use Tables for designers. Costs are expressed as pounds per unit of the element as shown. Elements are as defined in the Standard Form of Cost Analysis. All dimensions are given in millimetres unless otherwise stated.

5 On the right hand side are the underlying data such as cleaning frequencies, renewal and redecoration cycles, and the undiscounted costs of repairs etc on which the Costs-in-Use Tables are based.

6 In addition the mass and heat resistance of components are given on the right hand side to show the differences between one component and another when considering the effect of weight on the design and the achievement of U values and consequent heat savings.

Price Levels

7 The rates include overheads and profit and relate to the national average cost of materials, labour and plant current in 1st quarter 1991. A table showing the comparative differences for different locations is included in the memoranda. If the user wishes to add preliminaries to the figures, 12% is a reasonable average.

8 All rates are exclusive of VAT.

Discount Rate

9 Costs-in-Use are calculated over a building life of 60 years, discounted to 'present value' using a discount rate of 6%. This is the amount of money required to be invested now to provide the costs when shown. Tables of discount rates are given in the memoranda.

10 These costs have not been enhanced for inflation because the level of inflation of building costs over a long period is quite impossible to predict and, as a general rule, the true Costs-in-Use will be the same at any time during the life of the building since in the long run inflation in finance value tends to match inflation in building costs.

Capital Costs

11 In some instances it will be necessary to add two or more items together to obtain the complete cost of an element:

eg an external wall may comprise:

	Capital Cost	*Total Cost*
Plain clay tile hanging, battens and underfelt	36.49	37.54
100 blockwork outer skin, lightweight	16.61	16.62
Cavity with stainless steel ties	1.82	1.83
Cavity insulation 25 mm	3.64	3.65
100 blockwork inner skin, lightweight	16.61	16.62
Lightweight plaster	5.89	5.93
Internal decoration	1.89	10.95
	£82.95/m^2	£93.14/m^2

12 Since preliminaries are not included, this does not include scaffolding.

Cleaning Costs

13 The rates indicated in the tables relate to the cleaning only of the building element concerned as carried out by normal cleaning services. The external cleaning costs are restricted to window cleaning, and the cleaning of stonework or other surfaces which would deteriorate or become disfigured if not cleaned, eg aluminium window frames.

14 The rates do not include cleaning of a general nature (eg dusting furniture, clearing out waste paper baskets, changing towels, etc), which may be allowed for by a general rate of £1.15 (1990 prices) per m^2 of gross floor area per year.

Cyclical Costs

15 This category includes maintenance of a periodical and cyclical nature, the requirement for which is known to the designer when the choice of a component is made and which can vary depending on his selection. The rates included are based on recommendations made by manufacturers, clients, PSA maintenance staff etc.

16 Where components have an expected life of less than 60 years, renewal has been allowed at the original cost plus the cost of taking down the old work. Life expectancy is taken from a number of sources and relates to the effect of normal ageing processes or climatic conditions. No allowance is made here for any maintenance costs arising from likely obsolescence.

17 No allowance is made for the cost of day to day maintenance arising from normal accidental damage etc such as reglazing broken glass. But an allowance has been included within the Repairs Costs for repair of materials that are more prone to damage. General accidental damage repairs may be allowed for by a general rate of £1.90 (1990 prices) per m^2 of gross floor area per year.

18 The moving of partitions is an important factor in many buildings. This has been shown as a cost in this column and should only be included in the total if the client may require it.

19 Disturbance costs are often incurred by the occupants of the building during maintenance operations. The degree of disturbance can have a significant effect on productivity, particularly where rooms, or even entire buildings have to be temporarily vacated. No allowance has been made in the rates for likely disturbance costs. Where they may be significant they should always be taken into account by informing the client of the possible extent so that he may calculate his costs.

Repairs Costs

20 While it is expected that sound workmanship and materials will be used in conjunction with good design, there will inevitably be failures arising from unforeseeable causes and some forms of construction are more prone to failure than others. The costs shown will not therefore arise on every project on which a particular form of construction is used. The amounts shown are the costs of repair of the possible defects multiplied by the anticipated risk, so that if that construction is used on a number of projects the average cost of all of them will be the amount shown, even though most of them have no defects at all. The defects listed are only the most common ones that may arise.

21 In most cases the need to remedy defects has been predicted to arise at unforeseeable times within the stated period of years. The cost of remedial work has been spread as equal annual amounts over that period and discounted to present value.

Example calculation

22 *Plastic weather boarding to external wall.*

Repair physical damage £36 between years 1 and 60
÷ 60 = £0.60 p.a.
× 8% risk = £0.048 p.a.
present value of £1 p.a. @ 6% over 60 years
× 16.161 = £0.78

Mass

23 It should be noted that while the majority of weights are shown per m², there are exceptions where lineal or enumerated items occur. Weights can vary, particularly where the moisture content is allowed to change.

Heat Resistance (R value)

24 It is not possible to add directly the U values (W/m^2K) of the components of an element which is made up of several layers of materials. The R values (m^2K/W) can be added and hence these have been shown. If the user wishes to know the U value this can be calculated as the reciprocal of the R value.

25 The total R value of a component consists of the thermal resistance of that material plus the surface resistance. When a component comprises several layers of materials the surface resistance applies to the outer surfaces only. The surface resistance has not therefore been included in the R value shown but should be added to the total component value as shown in the example below.

26 In the case of windows and doors where the problem of adding further layers does not arise the R values shown *include* surface resistance.

Examples of calculation of U value

A *External Wall*

Outer skin, facing bricks, sandfaced	0.13
Cavity 50 mm (if 25 remains unfilled the value is the same)	0.18
Cavity insulation, 25 mm polystyrene	0.68
Inner skins, aerated blocks 100 mm	0.63
Lightweight plaster	0.08
Decoration	—
	1.70
Internal surface resistance (average)	0.12
External surface resistance (average)	0.06
Total thermal resistance R	1.88

$$\text{Thermal transmittance } U = \frac{1}{R} = \frac{1}{1.88} = 0.53$$

Notes i Internal surface resistance can be increased to 0.30 by use of a surface with low emissivity. This is governed by shape and type of surface*

ii External surface resistance varies from 0.03 to 0.11 with degree of exposure and height*

iii There are other factors which can affect the R and U values*

B *Window*

UPVC, double glazed, tilt and turn window	0.34
Surface resistances	INCLUDED
Total thermal resistance R	0.34

Thermal transmittance $U = \frac{1}{R} = \frac{1}{0.34} = 2.94$

* For further information see the CIBSE Guide A3 – Thermal Properties of Building Structures – 1980. Reprinted 1985.

27 The U values are used to determine annual energy costs. Energy costs also depend on the amount of heat required each year and the cost of fuel. Both these factors vary considerably. A general rate of £0.55 (1990 prices) multiplied by the U value will give an approximate allowance.

For example the annual heat loss through the external wall above will cost:

$0.53 \times 100m^2$ (area of wall) $\times$ £0.55 = £29.15

The Costs-in-Use over the 60 year life of the building discounted to present value at a rate of 6% will be:

£29.15 $\times$ 16.161 = £471.09

Generally

28 Little factual data exists on the lives of many materials and components, and in practice Costs-in-Use depend partly at least on factors peculiar to each building, such as the degree of exposure, the users of the buildings etc. A deliberately 'broad brush' approach has therefore been adopted, and minor details of specification have been ignored.

29 The costs and other data are based on the best information it has been possible to obtain within reasonable constraints of time and cost. Nevertheless they are inevitably based largely on informed opinion rather than hard fact. If Costs-in-Use are to form the basis of managerial and design decisions, and if the tables are to be extended to cover more forms of construction and minor details of specification, it is obviously important to improve the quality of the data. Any assistance which can be given to this end will be gratefully received and acknowledged.

Relevant information should be sent to:

Property Services Agency
Specialist Services
Room 1218
Apollo House
36 Wellesley Road
CROYDON
CR9 3RR

Tables

1 SUBSTRUCTURE

No	Description	Capital	Cleaning	Cyclical	Repairs	Total
	1.A.1 Piling					
1	80 piles 15m long to take 100 tonnes each; clear prepared site; cutting off tops; removing spoil where required.					
a	Driven steel piles	117,000.00	–	–	–	117,000.00
b	Driven reinforced concrete piles	41,000.00	–	–	–	41,000.00
c	Driven shell, cast in situ, reinforced concrete piles	58,000.00	–	–	–	58,000.00
d	Bored, cast in situ, reinforced concrete piles (i) Auger method (ii) Tripod method *Note:* Choice of method depends mainly on type of ground	 65,000.00 72,000.00	 – –	 – –	 – –	 65,000.00 72,000.00
	1.A.2 Wall Foundations					
1	STRIP FOUNDATION					
a	Excavating trench 1000 dp; disposal; earthwork support; 11.5N concrete foundation, 225; 265 cavity brickwork; damp proof course	83.00	–	–	0.06	83.06
2	TRENCH FILL FOUNDATION					
a	Excavating trench 1000 dp; disposal; earthwork support; 11.5N concrete fill; 265 cavity brickwork; damp proof course	64.00	–	–	0.06	64.06

Cleaning			Cyclical			Repairs					
Annual Cost	*Frequency No*	*Yr*	*Work*	*Cost*	*Cycle in Years*	*Cause/ Action*	*Cost (£) and Risk*	*Occurs in Years*	*kg/m²*	*R value*	*No*
											1
–	–	–	–	–	–	–	–	–	–		a
–	–	–	–	–	–				–	–	b
–	–	–	–	–	–	–	–	–	–	–	c
											d
–	–	–	–	–	–	–	–	–	–	–	(i)
–	–	–	–	–	–	–	–	–	–	–	(ii)
											1
–	–	–	–	–	–	Subsidence, under-pinning	300 0.1%	10–60	–	–	a
											2
–	–	–	–	–	–	Subsidence, under-pinning	300 0.1%	10–60	–	–	a

1 SUBSTRUCTURE

1.A.3 Ground floor structure

No	Description	Capital	Cleaning	Cyclical	Repairs	Total
1	SOLID GROUND FLOOR m^2					
a	Excavating to reduce levels 225 dp; disposal; 150 hardcore; 11.5N concrete bed, 150; mesh reinforcement 2.22kg/m^2; two coats Synthaprufe	21.49	–	–	0.19	21.68
b	Excavating to reduce levels 300dp; disposal; 150 hardcore; lean concrete blinding, 50; 25.5N concrete bed, 200; mesh reinforcement 3.02kg/m^2; 0.13 polythene damp proof membrane	27.87	–	–	0.26	28.13
c	*ADD* to items 1a and 1b above for					
	(i) excavating a further 150 and fill with hardcore	4.36	–	–	0.03	4.39
	(ii) insulation under slab, 50 expanded polystyrene	3.89	–	–	0.03	3.92
2	HOLLOW GROUND FLOOR m^2					
a	Excavating to reduce levels 375 dp; disposal; 100 hardcore; 11.5N concrete bed, 150; common brick sleeper wall, half brick, 300 high, honeycomb, 2m centres; 2 course slate dpc; 100 × 50 sw plates; 100 × 50 sw joists, 400 centres; all treated	30.33	–	–	0.24	30.57
b	*ADD* to item 2a above for					
	(i) excavating a further 150; additional 150 sleeper wall	3.08	–	–	0.03	3.11
	(ii) underfloor insulation					
	(1) 50 polystyrene between joists on battens	9.60	–	–	0.01	9.61
	(2) 25 quilt across joists	2.13	–	–	–	2.13
	(3) reinforced aluminium foil across joists	2.08	–	–	–	2.08
c	Excavating to reduce levels 375 dp; disposal; lean concrete blinding, 50; 0.13 polythene; 110 precast planks, 4m span; 50 reinforced screed	65.77	–	–	0.02	65.79
	For other spans see Element 2.B—'Upper Floors'					

Cleaning			Cyclical			Repairs					
Annual Cost	Frequency No	Yr	Work	Cost	Cycle in Years	Cause/ Action	Cost (£) and Risk	Occurs in Years	kg/m²	R value	No
											1
–	–	–	–	–	–	Subsidence, bad work/ materials, relay floor	26 1%	1–10	580	2.78	a
–	–	–	–	–	–	As above	36 1%	1–10	811	2.78	b
											c
–	–	–	–	–	–	As above	4 1%	1–10	225	–	(i)
–	–	–	–	–	–	As above	4 1%	1–10	1	0.36	(ii)
											2
–	–	–	–	–	–	Dry rot, replace timber & plaster	24 0.50%	10–60	598	2.70	a
						Subsidence, relay floor	30 1%	1–10			
											b
–	–	–	–	–	–	As above	4 1%	1–10	16	–	(i)
											(ii)
–	–	–	–	–	–	As above	10 0.50%	10–60	3	1.47	(1)
–	–	–	–	–	–	As above	2 0.50%	10–60	1	0.66	(2)
–	–	–	–	–	–	As above	2 0.50%	10–60	1	0.55	(3)
–	–	–	–	–	–	Defective work/ materials, replace floor	75 0.15%	10–60	491	2.70	c

2 SUPERSTRUCTURE

2.A Frame

No	Description	Capital	Cleaning	Cyclical	Repairs	Total
1	REINFORCED IN SITU CONCRETE(30N) m^2(GFA)					
a	Isolated columns at 3m centres; isolated beams between; 3m storey height	43.86	–	–	0.01	43.87
b	Isolated columns at 5m centres; isolated beams between; 3m storey height	61.61	–	–	0.02	61.63
c	Attached columns at 3m centres; attached beams between; 3m storey height	24.96	–	–	0.01	24.97
d	Attached columns at 5m centres; attached beams between; 3m storey height	42.61	–	–	0.01	42.62
2	SOFTWOOD, TREATED m^2(GFA)					
a	Studwork; plates, heads, noggings, braces, nail plates					
	(i) 50 × 100; 600 centres	12.67	–	–	–	12.67
	(ii) 50 × 100; 400 centres	15.06	–	–	0.01	15.07
	(iii) 75 × 100; 600 centres	17.47	–	–	0.01	17.48
	(iv) 75 × 100; 400 centres	20.86	–	–	0.01	20.87
3	STEELWORK m^2(GFA)					
a	Exposed; grade 43; UB's and RSC's primed;					
	(i) Columns 3m centres; beams between; 3m storey height	46.13	–	–	–	46.13
	(ii) Columns 5m centres; beams between; 3m storey height	59.62	–	–	–	59.62

Cleaning			Cyclical			Repairs					
Annual Cost	*Frequency No*	*Yr*	*Work*	*Cost*	*Cycle in Years*	*Cause/ Action*	*Cost (£) and Risk*	*Occurs in Years*	*kg/m²*	*R value*	*No*
											1
–	–	–	–	–	–	Defective work/ materials	47 0.15%	10–60	184	0.21 × 15% of area	a
–	–	–	–	–	–	As above	67 0.15%	10–60	276	0.32 × 26% of area	b
–	–	–	–	–	–	As above	28 0.15%	10–60	115	0.14 × 15% of area	c
–	–	–	–	–	–	As above	46 0.15%	10–60	207	0.25 × 26% of area	d
											2
											a
–	–	–	–	–	–	Dry rot/ woodworm	14 0.25%	20–60	10	0.08 × 18% of area	(i)
–	–	–	–	–	–	As above	18 0.25%	20–60	12	0.08 × 22% of area	(ii)
–	–	–	–	–	–	As above	21 0.25%	20–60	14	0.08 × 26% of area	(iii)
–	–	–	–	–	–	As above	24 0.25%	20–60	17	0.08 × 36% of area	(iv)
											3
											a
–	–	–	–	–	–	–	–	–	31	Ignore	(i)
–	–	–	–	–	–	–	–	–	42	Ignore	(ii)

2 SUPERSTRUCTURE

2.A Frame – *continued*

No	Description	Capital	Cleaning	Cyclical	Repairs	Total
3	STEELWORK – *continued* m²(GFA)					
b	Exposed; grade 43; RHS columns 3m centres; UB or RSC beams between; 3m storey height	49.57	–	–	–	49.57
c	Encased; grade 43; UB's and RSC's; concrete 50 cover					
	(i) Columns 3m centres; beams between; 3m storey height	81.87	–	–	0.01	81.88
	(ii) Columns 5m centres; beams between; 3m storey height	90.85	–	–	0.01	90.86

Cleaning			Cyclical			Repairs					
Annual Cost	Frequency No	Yr	Work	Cost	Cycle in Years	Cause/ Action	Cost (£) and Risk	Occurs in Years	kg/m^2	R value	No
											3
–	–	–	–	–	–	–	–	–	31	Ignore	b
											c
–	–	–	–	–	–	Replace defective casing	45 0.15%	10–60	260	0.23 × 29% of area	(i)
–	–	–	–	–	–	As above	36 0.15%	10–60	250	0.25 × 28% of area	(ii)

2 SUPERSTRUCTURE

2B Upper floors

No	Description	Capital	Cleaning	Cyclical	Repairs	Total
1	REINFORCED IN SITU CONCRETE m²					
a	21N concrete slab, 150 thick; hy reinforcement 20 kg/m²; formwork; 4m span	51.52	–	–	0.02	51.54
b	30N concrete slab, 225 thick; hy reinforcement 36kg/m²; formwork; 4.8m span	69.95	–	–	0.02	69.97
c	30N concrete slab, 225 thick; coffered or troughed; hy reinforcement 36 kg/m²; formwork; 4.8m span	61.68	–	–	0.02	61.70
2	PRECAST PRESTRESSED CONCRETE m²					
a	Units fixed direct from lorry; 1 hr fire grade; 50 reinforced screed					
	(i) 110 thick; 3m span	49.73	–	–	0.02	49.75
	(ii) 150 thick; 6m span	56.47	–	–	0.02	56.49
b	Hollow pots and precast beams; 150 thick; 50 reinforced screed; 6m span	35.09	–	–	0.01	35.10
3	SOFTWOOD, TREATED m²					
a	Joists, 400 centres; herring-bone strutting					
	(i) 175 × 50; 3m span	13.50	–	–	–	13.50
	(ii) 200 × 50; 4m span	14.83	–	–	0.01	14.84
	(iii) 225 × 50; 5m span	15.82	–	–	0.01	15.83

Cleaning			*Cyclical*			*Repairs*					
Annual Cost	*Frequency No*	*Yr*	*Work*	*Cost*	*Cycle in Years*	*Cause/ Action*	*Cost (£) and Risk*	*Occurs in Years*	*kg/m²*	*R value*	*No*
											1
–	–	–	–	–	–	Defective work/ materials	57 0.15%	10–60	345	0.11	a
–	–	–	–	–	–	As above	79 0.15%	10–60	518	0.16	b
–	–	–	–	–	–	As above	66 0.15%	10–60	276	0.09	c
											2
											a
–	–	–	–	–	–	Defective work/ materials	56 0.15%	10–60	368	0.11	(i)
–	–	–	–	–	–	As above	65 0.15%	10–60	460	0.14	(ii)
–	–	–	–	–	–	As above	41 0.15%	10–60	340	0.08	b
											3
											a
–	–	–	–	–	–	Dry rot/ woodworm	15 0.25%	20–60	14	0.33	(i)
–	–	–	–	–	–	As above	17 0.25%	20–60	16	0.35	(ii)
–	–	–	–	–	–	As above	18 0.25%	20–60	18	0.37	(iii)

2.C.1.A Flat roof structure

No	Description	Capital	Cleaning	Cyclical	Repairs	Total
1	REINFORCED IN SITU CONCRETE m²					
a	30N concrete slab, 200 thick; hy reinforcement, 30 kg/m²; formwork; 4m span	62.59	–	–	0.02	62.61
b	30N concrete slab, 200 thick; attached beams; 300 × 125; hy reinforcement, 30 kg/m² + 350 kg/m³; formwork; 5m span	76.38	–	–	0.02	76.40
c	30N concrete slab, 200 thick; hy reinforcement, 30 kg/m²; formwork; steel beams; reinforced casing; fabric beam wrapping; formwork, padstones; 5m span	81.83	–	–	0.03	81.86
d	30N concrete slab, troughed or coffered, 400 thick overall; fabric reinforcement; hy reinforcement to ribs; formwork; 5m span	60.03	–	–	0.02	60.05
2	SOFTWOOD/COMBINATION m²					
a	Sawn softwood joists; 50 × 100, 400 centres; solid strutting; plate to perimeter; steel beams at 1200 centres; padstones; 5m span	26.91	–	–	0.03	26.94
b	Glued laminated beams, 115 × 250, 1.5m centres; sawn softwood bearers, 50 × 75, 600 centres; joist hangers; fixing cramps; 4m span	29.46	–	–	0.09	29.55
c	Glued laminated beams, 115 × 400, 600 centres; top edges secured by deck; 6m span	55.31	–	–	0.17	55.48
d	Plywood/softwood combination beams, 100 × 305, 1200 centres; 50 × 50 noggings, 600 centres; 6m span	25.56	–	–	0.08	25.64
e	Corrugated steel/softwood combination beams, 38 × 64 × 226, 600 centres; 50 × 50 softwood noggings/softwood perimeter plate; 3m span	74.86	–	–	0.21	75.07
f	Corrugated steel/softwood combination beams, 38 × 89 × 326, 600 centres; 50 × 50 softwood noggings/softwood perimeter plate; 6m span	89.27	–	–	0.26	89.53

Cleaning			Cyclical			Repairs					
Annual Cost	*Frequency No*	*Yr*	*Work*	*Cost*	*Cycle in Years*	*Cause/ Action*	*Cost (£) and Risk*	*Occurs in Years*	*kg/m²*	*R value*	*No*
											1
–	–	–	–	–	–	Defective work/ materials	70 0.15%	10–60	460	0.14	a
–	–	–	–	–	–	As above	85 0.15%	10–60	546	0.18	b
–	–	–	–	–	–	As above	91 0.15%	10–60	532	0.16	c
–	–	–	–	–	–	As above	67 0.15%	10–60	414	0.13	d
											2
–	–	–	–	–	–	Dry rot, woodworm	13 1%	10–40	30	0.18	a
–	–	–	–	–	–	Dry rot, woodworm, delamination	33 1%	10–40	21	0.24	b
–	–	–	–	–	–	Delamination	64 1%	10–40	5	0.20	c
–	–	–	–	–	–	Defective materials, dry rot, woodworm	31 1%	10–40	10	0.20	d
–	–	–	–	–	–	Defective materials, corrosion, dry rot, woodworm	78 1%	10–40	9	0.17	e
–	–	–	–	–	–	As above	96 1%	10–40	10	0.16	f

2.C.1.A Flat roof structure – *continued*

No	*Description*		*Capital*	*Cleaning*	*Cyclical*	*Repairs*	*Total*
3	STEEL	m^2					
a	Galvanised steel, Z profile purlins; 1m centres; fixing by cleats; 6m span		9.79	–	–	–	9.79
b	Steel prefabricated lattice joists, 600 centres; softwood top and bottom fixings; ends built in; top edge secured by deck						
	(i) 200 deep; 5m span		54.27	–	–	0.15	54.42
	(ii) 300 deep; 7m span		63.38	–	–	0.18	63.56
	(iii) 450 deep; 10m span		90.68	–	–	0.27	90.95
4	*ADD* where there is to be no ceiling, for decoration and cleaning	m^2					
	(i) Items 1a–d — see Element 3C internal finishings; ceiling decoration		–	–	–	–	–
	(ii) Items 2a, b, d — Beams/trusses 1.2 to 1.5m apart		5.24	0.65	8.39	–	14.28
	(iii) Items 2c, e, f, 3b — Beams/trusses 600 apart		5.12	0.65	8.21	–	13.98
	(iv) Items 3a — Purlins		2.46	2.75	3.96	–	9.17
5	DECKING	m^2					
a	Sawn softwood boards, t & g, 19 thick		10.86	–	–	0.03	10.89
b	Plywood WBP, t & g, 18 thick		14.40	–	–	0.05	14.45
c	Chipboard, t & g, moisture reistant, 18 thick		6.27	–	–	0.04	6.31

Cleaning			*Cyclical*			*Repairs*					
Annual Cost	*Frequency No*	*Yr*	*Work*	*Cost*	*Cycle in Years*	*Cause/ Action*	*Cost (£) and Risk*	*Occurs in Years*	*kg/m²*	*R value*	*No*
											3
–	–	–	–	–	–	Corrosion	11 0.25%	20–60	5	–	a
											b
–	–	–	–	–	–	Corrosion of steel, rot of timber	58 1%	10–40	12	0.18	(i)
–	–	–	–	–	–	As above	68 1%	10–40	15	0.18	(ii)
–	–	–	–	–	–	As above	103 1%	10–40	30	0.18	(iii)
											4
–	–	–	–	–	–	–	–	–	–	–	(i)
Dust 0.04	1	1	Repaint	3.67	6	–	–	–	–	–	(ii)
Dust 0.04	1	1	Repaint	3.59	6	–	–	–	–	–	(iii)
Dust 0.17	1	1	Repaint	1.73	6	–	–	–	–	–	(iv)
											5
–	–	–	–	–	–	Dry rot, woodworm	13 1%	10–40	10	0.15	a
–	–	–	–	–	–	Defective work/ materials	17 1%	10–40	10	0.15	b
–	–	–	–	–	–	Damp	8 2%	10–40	11	0.15	c

2.C.1.A Flat roof structure — *continued*

No	Description	Capital	Cleaning	Cyclical	Repairs	Total
5	DECKING — *continued* m²					
d	Plywood exterior grade, 8 thick; urethane foam, aluminium foil, combination board					
	(i) 25 thick	17.34	–	–	0.05	17.39
	(ii) 50 thick	21.22	–	–	0.06	21.28
e	Woodwool, standard finish, 50 thick	9.13	–	–	0.04	9.17
f	Woodwool, prescreeded, 50 thick	10.20	–	–	0.05	10.25
g	Woodwool, steel edged, 50 thick, textured soffite; 2.1m span	19.31	–	–	0.09	19.40
h	Strawboard, 50 thick; joints taped	11.44	–	–	0.07	11.51
j	Galvanised steel, troughed, 0.7 thick; bolted seams					
	(i) 48 deep troughs; 1.5m span	17.06	–	–	0.01	17.07
	(ii) 100 deep troughs; 3.5m span	20.65	–	–	0.01	20.66
k	Aluminium, troughed, 0.91 thick; bolted seams					
	(i) 63 deep troughs; 1.5m span	23.35	–	–	0.02	23.37
	(ii) 100 deep troughs; 2.9m span	26.06	–	–	0.02	26.08

Cleaning			Cyclical			Repairs					
Annual Cost	Frequency No	Yr	Work	Cost	Cycle in Years	Cause/ Action	Cost (£) and Risk	Occurs in Years	kg/m^2	R value	No
											5
											d
–	–	–	–	–	–	Defective materials	19 1%	10–40	6	0.98	(i)
–	–	–	–	–	–	As above	23 1%	10–40	6	2.12	(ii)
–	–	–	–	–	–	As above	11 1.50%	10–40	25	0.50	e
–	–	–	–	–	–	As above	12 1.50%	10–40	28	0.48	f
–	–	–	–	–	–	As above	22 1.50%	10–40	33	0.44	g
–	–	–	–	–	–	Defective materials, damp	13 2%	10–40	18	0.46	h
											j
–	–	–	–	–	–	Corrosion	19 0.50%	20–60	8	0.09	(i)
–	–	–	–	–	–	As above	23 0.50%	20–60	11	0.09	(ii)
											k
–	–	–	–	–	–	As above	25 0.50%	20–60	4	0.09	(i)
–	–	–	–	–	–	As above	28 0.50%	20–60	5	0.09	(ii)

2.C.1.A Flat roof structure — *continued*

No	Description	Capital	Cleaning	Cyclical	Repairs	Total
6	INSULATION m²					
	ADD to above where appropriate including vapour barrier and "adhesive"			*Add* 20% if under felt coverings	*Add* 20% if under felt coverings	
a	Lightweight screed					
	(i) 25 thick	11.57	–	2.31	0.32	14.20
	(ii) 60 average thick	14.93	–	2.98	0.40	18.31
b	Polyurethene or Polyisocyanurate board					
	(i) 35 thick	13.83	–	2.76	0.38	16.97
	(ii) 50 thick	15.91	–	3.18	0.40	19.49
c	Glass fibre board 60 thick	12.49	–	2.50	0.32	15.31
d	Fibre board, bitumen impregnated 12.7 thick	9.50	–	1.90	0.27	11.67
e	Polystyrene expanded board					
	(i) 50 thick	13.04	–	2.61	0.35	16.00
	(ii) 75 thick	15.66	–	3.13	0.40	19.19
f	Glass foamed cellular board 60 thick	21.02	–	4.20	0.57	25.79
g	Perlite expanded granule slab 60 thick	17.31	–	3.46	0.48	21.25

Cleaning			Cyclical			Repairs					
Annual Cost	*Frequency No*	*Yr*	*Work*	*Cost*	*Cycle in Years*	*Cause/ Action*	*Cost (£) and Risk*	*Occurs in Years*	*kg/m²*	*R value*	*No*
											6
											a
–	–	–	Repair/ replace when replacing felt coverings	5.65	20	Repair if under felt coverings	12 10%	1–60	23	0.16	(i)
–	–	–	As above	7.29	20	As above	15 10%	1–60	54	0.38	(ii)
											b
–	–	–	As above	6.75	20	As above	14 10%	1–60	1	1.40	(i)
–	–	–	As above	7.78	20	As above	15 10%	1–60	2	2.00	(ii)
–	–	–	As above	6.11	20	As above	12 10%	1–60	2	1.71	c
–	–	–	As above	4.64	20	As above	10 10%	1–60	4	0.25	d
											e
–	–	–	As above	6.37	20	As above	13 10%	1–60	1	1.43	(i)
–	–	–	As above	7.66	20	As above	15 10%	1–60	2	2.14	(ii)
–	–	–	As above	10.27	20	As above	21 10%	1–60	8	1.00	f
–	–	–	As above	8.47	20	As above	18 10%	1–60	11	1.02	g

2.C.2.A Flat roof coverings

No	Description		Capital	Cleaning	Cyclical	Repairs	Total
1	SINGLE SHEET	m^2					
a	Lead sheet, rolls, clips, drips; building paper underlay						
	(i) Code 4, 1.8 thick		54.54	–	–	0.07	54.61
	(ii) Code 5, 2.24 thick		63.65	–	–	0.08	63.73
	(iii) Code 6, 2.5 thick		71.89	–	–	0.09	71.98
b	Zinc sheet, seams, clips; building paper underlay						
	(i) 12 gauge, 0.65 thick		42.65	–	–	0.06	42.71
	(ii) 14 gauge, 0.81 thick		45.95	–	–	0.07	46.02
c	Copper sheet, seams, clips; building paper underlay						
	(i) 24 gauge, 0.55 thick		58.27	–	–	0.12	58.39
	(ii) 22 gauge, 0.7 thick		70.65	–	–	0.15	70.80
d	Aluminium sheet, seams, clips; building paper underlay 0.9 thick		51.10	–	–	0.13	51.23
2	ASPHALT	m^2					
a	Asphalt, two coat, 20; felt underlay						
	(i) BS 988, limestone aggregate		12.62	–	–	0.21	12.83
	(ii) BS 6577, rock asphalt aggregate		16.62	–	–	0.26	16.88

Cleaning			Cyclical			Repairs					
Annual Cost	Frequency No	Yr	Work	Cost	Cycle in Years	Cause/ Action	Cost (£) and Risk	Occurs in Years	kg/m²	R value	No
											1
											a
–	–	–	–	–	–	Defective work	57 0.20%	1–20	21	ignore	(i)
–	–	–	–	–	–	As above	66 0.20%	1–20	26	ignore	(ii)
–	–	–	–	–	–	As above	75 0.20%	1–20	29	ignore	(iii)
											b
–	–	–	–	–	–	Defective work	45 0.20%	1–20	5	ignore	(i)
						Corrosion	45 0.20%	20–60			
–	–	–	–	–	–	As above	48 0.20%	1–20	6	ignore	(ii)
							48 0.20%	20–60			
											c
–	–	–	–	–	–	As above	61 0.20%	1–20	5	ignore	(i)
							61 1%	30–60			
–	–	–	–	–	–	As above	74 0.20%	1–20	6	ignore	(ii)
							74 1%	30–60			
											d
–	–	–	–	–	–	As above	54 0.20%	1–20	3	ignore	
							54 1%	20–60			
											2
											a
–	–	–	–	–	–	Cracks and blisters	17 10%	20–60	36	0.05	(i)
–	–	–	–	–	–	As above	21 10%	20–60	40	0.04	(ii)

No	2.C.2.A **Flat roof coverings** – *continued* Description	Capital	Cleaning	Cyclical	Repairs	Total
3	BUILT UP LAYERS m²					
a	Bituminous felt, BS 747; two layer, type 3, bonded	7.89	–	5.90	0.48	14.27
b	Bituminous felt, BS 747; three layer, type 2B, bonded	13.74	–	6.02	0.75	20.51
c	High performance; glass fibre polyester base bituminous 1.8kg/m², two layer, bonded	9.70	–	7.10	0.59	17.39
d	High performance; glass fibre polyester base bituminous; one layer 1.8kg/m², one layer 4.25kg/m², bonded	14.60	–	8.04	0.61	23.25
e	Elastomeric; polyester base bituminous; one layer 1.8kg/m², one layer 4 kg/m², bonded	13.76	–	6.03	0.57	20.36
f	Elastomeric; polyester/glass fibre base bituminous, one layer 1.8kg/m², one layer 2kg/m², bonded	13.48	–	4.15	0.57	18.20
g	Elastomeric; polyester/glass fibre base bituminous; base layer, one layer, 1.8kg/m², bonded	19.05	–	3.49	0.54	23.08
h	Metal faced, glass fibre based bituminous, one layer 1.8kg/m², one layer copper faced 4.72kg/m² bonded	24.26	–	4.39	0.34	28.99
j	Laminated bitumen/fibre semi-rigid sheets; single layer 2.45kg/m², jointing strips, bonded	15.54	–	4.76	0.46	20.74
4	SUNDRIES m² *ADD* to the above where appropriate					
a	Perforated underlay for venting, partially bonded	4.21	–	1.72	0.11	6.04
b	Chippings as reflective layer	3.37	–	1.63	0.11	5.11
c	Solar reflective paint	2.09	–	2.75	0.05	4.89
d	Mineral finish to top layer in lieu of sanded finish	0.86	–	0.35	0.03	1.24
e	Terrace tiles, mineral cement, 300 × 300 × 12, bedded in hot bitumen	30.64	–	3.15	0.89	34.68

Cleaning			Cyclical			Repairs					
Annual Cost	*Frequency No*	*Yr*	*Work*	*Cost*	*Cycle in Years*	*Cause/ Action*	*Cost (£) and Risk*	*Occurs in Years*	*kg/m²*	*R value*	*No*
											3
–	–	–	Replace	8.88	15	Repair	9 20%	1–60	5	0.01	a
–	–	–	As above	14.73	20	As above	14 20%	1–60	6	0.02	b
–	–	–	As above	10.69	15	As above	11 20%	1–60	5	0.01	c
–	–	–	As above	15.59	18	As above	15 15%	1–60	8	0.02	d
–	–	–	As above	14.75	20	As above	14 15%	1–60	7	0.01	e
–	–	–	As above	14.47	25	As above	14 15%	1–60	6	0.01	f
–	–	–	As above	20.04	30	As above	20 10%	1–60	8	0.02	g
–	–	–	As above	25.25	30	As above	25 5%	1–60	8	0.01	h
–	–	–	As above	16.53	25	As above	17 10%	1–60	3	0.01	j
											4
–	–	–	As above	4.21	20	–	4 10%	1–60	2	0.01	a
–	–	–	As above	3.99	20	–	4 10%	1–60	18	0.01	b
–	–	–	Repaint	2.30	10	–	2 10%	1–60	–	ignore	c
–	–	–	Replace	0.86	20	–	1 10%	1–60	1	0.01	d
–	–	–	As above	32.51	40	Refix loose tiles	33 10%	1–60	22	0.03	e

2.C.1.B Pitched roof structure

No	Description	Capital	Cleaning	Cyclical	Repairs	Total
1	SOFTWOOD, TREATED m² ON PLAN					
a	Traditional construction; 100 × 50 rafters and ceiling joists, 375 centres; 150 × 50 purlins; plates; ridge; hangers					
	(i) 30° pitch	21.35	–	–	0.02	21.37
	(ii) 35° pitch	22.23	–	–	0.02	22.25
	(iii) 40° pitch	23.16	–	–	0.02	23.18
b	Gangnail trusses, 1.8m centres; 100 × 38 rafters and ceiling joists, 450 centres; 150 × 50 purlins; plates; ridge; hangers					
	(i) 30° pitch	16.71	–	–	0.01	16.72
	(ii) 35° pitch	17.22	–	–	0.01	17.23
	(iii) 40° pitch	17.85	–	–	0.01	17.86
c	Traditional trusses, 600 centres; 75 × 50 rafters and ceiling joists; plates; ridge; struts; 20° pitch	12.45	–	–	0.01	12.46
2	STEEL m² ON PLAN					
a	Framed trusses; angle members bolted; angle purlins; galvanising; priming and painting two coats; padstones					
	(i) 30° pitch; 3m centres; 6 to 12m span	38.21	0.32	5.33	0.01	43.87
	(ii) 22½° pitch; 3.8m centres; 12 to 18m span	44.24	0.32	6.01	0.01	50.58

Cleaning			*Cyclical*			*Repairs*					
Annual Cost	*Frequency No*	*Yr*	*Work*	*Cost*	*Cycle in Years*	*Cause/ Action*	*Cost (£) and Risk*	*Occurs in Years*	*kg/m²*	*R value*	*No*
											1
											a
–	–	–	–	–	–	Dry rot/ wet rot/ woodworm	26 0.50%	20–60	21	–	(i)
–	–	–	–	–	–	As above	28 0.50%	20–60	22	–	(ii)
–	–	–	–	–	–	As above	28 0.50%	20–60	23	–	(iii)
											b
–	–	–	–	–	–	As above	20 0.50%	20–60	15	–	(i)
–	–	–	–	–	–	As above	21 0.50%	20–60	16	–	(ii)
–	–	–	–	–	–	As above	21 0.50%	20–60	16	–	(iii)
–	–	–	–	–	–	As above	14 0.50%	20–60	10	–	c
											2
											a
Dust 0.02	2	1	Repaint	2.33	6	Corrosion	42 0.50%	40–60	20	–	(i)
As above 0.02	2	1	As above	2.63	6	As above	48 0.50%	40–60	23	–	(ii)

2.C.1.B Pitched roof structure – *continued*

No	Description	Capital	Cleaning	Cyclical	Repairs	Total
3	DECKING, SLOPING m² ON PLAN					
a	Sawn softwood boards, t & g, 19 thick	14.98	–	–	0.01	14.99
b	Plywood WBP, t & g, 18 thick	16.47	–	–	0.02	16.49
c	Chipboard, t & g, moisture resistant, 18 thick	9.31	–	–	0.02	9.33
d	Galvanised steel, troughed, 0.7 thick; bolted seams					
	(i) 48 deep troughs; 1.5m span	18.78	–	–	0.01	18.79
	(ii) 100 deep troughs; 3.5m span	22.70	–	–	0.02	22.72
e	Alumimium, troughed, 0.91 thick; bolted seams					
	(i) 63 deep troughs; 1.5m span	25.69	–	–	0.02	25.71
	(ii) 100 deep troughs; 2.9m span	28.67	–	–	0.02	28.69
4	INSULATION m² ON PLAN *ADD* to above where appropriate					
a	Polystyrene expanded board, 50 thick	7.06	–	4.94	0.22	12.22
b	Glassfibre mat infill, 100 thick	4.79	–	–	–	4.79
c	Mineral wool quilt, laid, 100 thick	3.47	–	–	–	3.47

Cleaning			Cyclical			Repairs					
Annual Cost	Frequency No	Yr	Work	Cost	Cycle in Years	Cause/ Action	Cost (£) and Risk	Occurs in Years	kg/m²	R value	No
											3
–	–	–	–	–	–	Dry rot, woodworm	17 1%	30–60	11	0.16	a
–	–	–	–	–	–	Defective materials	19 1%	30–60	11	0.15	b
–	–	–	–	–	–	Damp	10 2%	20–60	12	0.16	c
											d
–	–	–	–	–	–	Corrosion	24 0.50%	20–60	9	0.09	(i)
–	–	–	–	–	–	As above	28 0.50%	20–60	12	0.09	(ii)
											e
–	–	–	–	–	–	As above	31 0.50%	20–60	4	0.09	(i)
–	–	–	–	–	–	As above	34 0.50%	20–60	5	0.09	(ii)
											4
–	–	–	Repair/replace when replacing felt coverings	7.44	15	Repair if under felt coverings	8 10%	1–60	2	1.71	a
–	–	–	–	–	–	–	–	–	6	2.81	b
–	–	–	–	–	–	–	–	–	20	2.68	c

2.C.2.B Pitched roof coverings

No	Description	Capital	Cleaning	Cyclical	Repairs	Total
1	PROFILED SHEETING m² ON PLAN					
a	Fibre cement profiled, natural colour					
	(i) to timber	12.50	–	1.41	0.08	13.99
	(ii) to steel	13.68	–	1.52	0.08	15.28
b	Fibre cement profiled, acrylic colour coated					
	(i) to timber	13.71	–	1.52	0.08	15.31
	(ii) to steel	14.91	–	1.64	0.09	16.64
c	Fibre cement profiled, natural colour, sandwich; bonded to insulation 60 thick; to steel	27.58	–	2.92	0.16	30.66
d	Steel profiled, PVC colour coated, 0.7 thick					
	(i) to timber	16.63	–	3.28	0.08	19.99
	(ii) to steel	18.16	–	3.58	0.08	21.82
e	Aluminium profiled, anodised or pre-coated, 0.7 thick					
	(i) to timber	21.10	–	4.16	0.09	25.35
	(ii) to steel	22.61	–	4.46	0.10	27.17
f	G.R.P. profiled, translucent, 1.6 thick; to timber	16.37	–	12.19	0.41	28.97

Cleaning			Cyclical			Repairs					
Annual Cost	Frequency No	Yr	Work	Cost	Cycle in Years	Cause/ Action	Cost (£) and Risk	Occurs in Years	kg/m²	R value	No
											1
											a
–	–	–	Renew	14.49	40	Wind damage, defective materials	14 2%	1–60	10	0.03	(i)
–	–	–	As above	15.68	40	As above	15 2%	1–60	10	0.03	(ii)
											b
–	–	–	As above	15.70	40	As above	15 2%	1–60	10	0.04	(i)
–	–	–	As above	16.90	40	As above	17 2%	1–60	10	0.04	(ii)
–	–	–	As above	30.11	40	As above	30 2%	1–60	23	2.76	c
											d
–	–	–	As above	18.84	30	As above	19 1.50%	1–60	7	ignore	(i)
–	–	–	As above	20.57	30	As above	20 1.50%	1–60	7	ignore	(ii)
											e
–	–	–	As above	23.90	30	As above	23 1.50%	1–60	3	ignore	(i)
–	–	–	As above	25.61	30	As above	24 1.50%	1–60	3	ignore	(ii)
–	–	–	As above	18.36	15	As above	19 8%	1–60	4	ignore	f

2.C.2.B Pitched roof coverings — *continued*

No	*Description*	*Capital*	*Cleaning*	*Cyclical*	*Repairs*	*Total*
2	SLATING AND TILING m² ON PLAN					
a	Clay plain tiles, 510 × 255, 65 lap; softwood battens; felt underlay					
	(i) machine made (P.C. £200/K)	42.15	–	–	1.21	43.36
	(ii) hand made (P.C. £500/K)	67.76	–	–	1.89	69.65
b	Clay interlocking tiles, 610 × 305, 75 lap; softwood battens; felt underlay					
	(i) low profile (P.C. £700/K)	29.05	–	–	0.86	29.91
	(ii) high profile (P.C. £1200/K)	48.85	–	–	1.40	50.25
c	Concrete plain tiles (P.C. £200/K) 267 × 165, 65 lap; softwood battens; felt underlay	40.63	–	–	1.16	41.79
d	Concrete interlocking tiles, 381 × 229, 75 lap; softwood battens; felt underlay					
	(i) low profile (P.C. £350/K)	13.02	–	–	0.40	13.42
	(ii) medium profile (P.C. £500/K)	16.35	–	–	0.51	16.86
	(iii) high profile (P.C. £650/K)	20.45	–	–	0.62	21.07
e	Natural slates, Welsh blue, 75 lap; softwood battens; felt underlay					
	(i) 510 × 255	44.61	–	–	1.27	45.88
	(ii) 610 × 305	45.49	–	–	1.29	46.78
f	Reconstituted slates, 300 × 336, interlocking, Welsh blue, 50 lap; softwood battens; felt underlay	30.01	–	–	0.89	30.90
g	Reconstituted stone slabs, random sizes, Cotswold/Moordale (P.C. £15/m²), 80 lap; softwood battens; felt underlay	42.64	–	–	1.21	43.85

Cleaning			Cyclical			Repairs					
Annual Cost	*Frequency No*	*Yr*	*Work*	*Cost*	*Cycle in Years*	*Cause/ Action*	*Cost (£) and Risk*	*Occurs in Years*	*kg/m²*	*R value*	*No*
											2
											a
–	–	–	–	–	–	Wind damage, defective materials	45 10%	1–60	81	0.27	(i)
–	–	–	–	–	–	As above	70 10%	1–60	81	0.27	(ii)
											b
–	–	–	–	–	–	As above	32 10%	1–60	72	0.26	(i)
–	–	–	–	–	–	As above	52 10%	1–60	72	0.26	(ii)
–	–	–	–	–	–	As above	43 10%	1–60	114	0.27	c
											d
–	–	–	–	–	–	As above	15 10%	1–60	89	0.20	(i)
–	–	–	–	–	–	As above	19 10%	1–60	89	0.20	(ii)
–	–	–	–	–	–	As above	23 10%	1–60	89	0.20	(iii)
											e
–	–	–	–	–	–	As above	47 10%	1–60	77	0.23	(i)
–	–	–	–	–	–	As above	48 10%	1–60	77	0.23	(ii)
–	–	–	–	–	–	As above	33 10%	1–60	82	0.23	f
–	–	–	–	–	–	As above	45 10%	1–60	137	0.27	g

2.C.2.B Pitched roof coverings — *continued*

No	Description	Capital	Cleaning	Cyclical	Repairs	Total
2	SLATING AND TILING — *continued* m² ON PLAN					
h	Slates, fibre cement, 600 × 300 or 500 × 250, 75 lap; softwood battens; felt underlay					
	(i) blue black/standard quality	31.72	–	–	0.92	32.64
	(ii) coloured/superior quality	33.97	–	–	0.97	34.94
j	Glassfibre reinforced bitumen strip, replica slates, mineral finish, 155 gauge, nailed to timber, edges torched; felt underlay	24.49	–	4.46	0.62	29.57
3	SHEET COVERINGS m² ON PLAN					
a	Lead sheet, rolls, drips; building paper underlay					
	(i) Code 4, 1.8 thick	63.22	–	–	0.09	63.31
	(ii) Code 5, 2.24 thick	74.28	–	–	0.10	74.38
	(iii) Code 6, 2.5 thick	82.05	–	–	0.11	82.16
b	Zinc sheet, seams, clips; building paper underlay					
	(i) 12 gauge, 0.65 thick	51.63	–	–	0.23	51.86
	(ii) 14 gauge, 0.8 thick	57.84	–	–	0.26	58.10
c	Copper sheet, seams, clips; felt underlay					
	(i) 24 gauge, 0.55 thick	74.05	–	–	0.16	74.21
	(ii) 23 gauge, 0.7 thick	85.67	–	–	0.21	85.88

Cleaning			Cyclical			Repairs					
Annual Cost	*Frequency No*	*Yr*	*Work*	*Cost*	*Cycle in Years*	*Cause/ Action*	*Cost (£) and Risk*	*Occurs in Years*	*kg/m²*	*R value*	*No*
											h
–	–	–	–	–	–	Wind damage, defective materials	34 10%	1–60	20	0.54	(i)
–	–	–	–	–	–	As above	36 10%	1–60	20	0.54	(ii)
–	–	–	Renew	25.62	30	As above	23 10%	1–60	13	0.06	j
											3
											a
–	–	–	–	–	–	Defective work	78 0.20%	1–20	23	ignore	(i)
–	–	–	–	–	–	As above	90 0.20%	1–20	29	ignore	(ii)
–	–	–	–	–	–	As above	99 0.20%	1–20	34	ignore	(iii)
											b
–	–	–	–	–	–	Defective work	65 0.20%	1–20	5	ignore	(i)
						Corrosion	65 2%	20–60			
–	–	–	–	–	–	As above	72 0.20%	1–20	7	ignore	(ii)
							72 2%	20–60			
											c
–	–	–	–	–	–	As above	89 0.20%	1–20	7	ignore	(i)
							72 1%	30–60			
–	–	–	–	–	–	As above	102 0.20%	1–20	9	ignore	(ii)
							102 1%	30–60			

2.C.2.B Pitched roof coverings — *continued*

No	Description	Capital	Cleaning	Cyclical	Repairs	Total
3	SHEET COVERINGS — *continued* m^2 ON PLAN					
d	Aluminium sheet, seams, clips, polythene and felt underlay, 0.9 thick	67.16	–	–	0.19	67.35
4	BUILT UP LAYERS m^2 ON PLAN					
a	Bituminous felt, BS 747; one layer type 2B, one layer type 2E, bonded	13.98	–	6.18	0.75	20.91
b	Bituminous felt, BS 747, two layers type 2B, one layer type 2E, bonded	19.82	–	8.57	1.13	29.52
c	Laminated bitumen/fibre semi-rigid sheets, single layer 2.45 kg/m^2; jointing strips; bonded	16.76	–	5.14	0.48	22.38

Cleaning			Cyclical			Repairs					
Annual Cost	*Frequency No*	*Yr*	*Work*	*Cost*	*Cycle in Years*	*Cause/ Action*	*Cost (£) and Risk*	*Occurs in Years*	*kg/m²*	*R value*	*No*
											3
–	–	–	–	–	–	Defective work Corrosion	81 0.20% 81 1%	1–20 20–60	5	ignore	d
											4
–	–	–	Renew	15.11	20	Repair	14 20%	1–60	6	0.02	a
–	–	–	As above	20.96	20	As above	21 20%	1–60	9	0.04	b
–	–	–	As above	17.90	25	As above	18 10%	1–60	3	0.01	c

No	Description		Capital	Cleaning	Cyclical	Repairs	Total
	2.C.3 Roof drainage						
1	RAINWATER PIPES	m					
a	Aluminium, BS 2997, ears cast on, polyester powder coated; plugged and nailed						
	(i) 63 nominal		11.44	–	2.08	0.33	13.85
	(ii) 75 nominal		13.64	–	2.08	0.40	16.12
	(iii) 100 nominal		19.60	–	2.08	0.53	22.21
b	Cast iron, BS 460, ears cast on, plugged and nailed; priming, painting two coats						
	(i) 50 nominal		16.90	–	3.86	0.46	21.22
	(ii) 63 nominal		17.55	–	3.86	0.49	21.90
	(iii) 75 nominal		18.56	–	3.86	0.51	22.93
	(iv) 100 nominal		23.20	–	3.86	0.62	27.68
	(v) 100 × 75 rectangular		62.78	–	3.86	1.46	68.10
c	UPVC, BS 4576, fixing brackets, plugged and screwed						
	(i) 68 nominal		5.81	–	–	0.22	6.03
	(ii) 110 nominal		9.48	–	–	0.29	9.77
	(iii) 68.5 square		5.26	–	–	0.22	5.48

Cleaning			Cyclical			Repairs					
Annual Cost	*Frequency No*	*Yr*	*Work*	*Cost*	*Cycle in Years*	*Cause/ Action*	*Cost (£) and Risk*	*Occurs in Years*	*kg/m*	*R value*	*No*
											1
											a
–	–	–	Repaint	1.36	5#	Defective work/ materials	15 3%	1–10	2	–	(i)
–	–	–	As above	1.36	5#	As above	18 3%	1–10	2	–	(ii)
–	–	–	As above	1.36	5#	As above	24 3%	1–10	3	–	(iii)
											b
–	–	–	As above	1.36	5	As above	21 3%	1–10	6	–	(i)
–	–	–	As above	1.36	5	As above	22 3%	1–10	7	–	(ii)
–	–	–	As above	1.36	5	As above	23 3%	1–10	10	–	(iii)
–	–	–	As above	1.36	5	As above	28 3%	1–10	16	–	(iv)
–	–	–	As above	1.36	5	As above	66 3%	1–10	20	–	(v)
											c
–	–	–	–	–	–	As above	10 3%	1–10	1	–	(i)
–	–	–	–	–	–	As above	13 3%	1–10	2	–	(ii)
–	–	–	–	–	–	As above	10 3%	1–10	1	–	(iii)

Commencing year 15

2.C.3 Roof drainage — *continued*

No	Description	Capital	Cleaning	Cyclical	Repairs	Total
2	GUTTERS m					
a	Aluminium, BS 2997, polyester powder coated, brackets screwed to timber					
	(i) 100 half round	11.40	3.23	4.50	0.38	19.51
	(ii) 112 half round	11.91	3.23	4.50	0.40	20.04
	(iii) 125 half round	14.32	3.23	4.50	0.44	22.49
	(iv) 100 ogee	13.27	3.23	4.50	0.44	21.44
	(v) 112 ogee	14.74	3.23	4.50	0.44	22.91
	(vi) 125 ogee	16.42	3.23	4.50	0.49	24.64
b	Cast iron, BS 460, brackets screwed to timber; priming, painting two coats					
	(i) 100 half round	13.08	3.23	8.34	0.42	25.07
	(ii) 125 half round	15.43	3.23	8.65	0.46	27.77
	(iii) 150 half round	21.96	3.23	9.79	0.62	35.60
	(iv) 100 ogee	13.65	3.23	8.65	0.42	25.95
	(v) 125 ogee	15.80	3.23	9.79	0.46	29.28
c	UPVC, BS 4576, brackets screwed to timber					
	(i) 76 half round	5.12	3.23	–	0.22	8.57
	(ii) 112 half round	6.00	3.23	–	0.24	9.47

Cleaning			Cyclical			Repairs					
Annual Cost	Frequency No	Yr	Work	Cost	Cycle in Years	Cause/ Action	Cost (£) and Risk	Occurs in Years	kg/m	R value	No
											2
											a
Clean out and wash 0.20	1	1	Repaint	2.94	5#	Defective work/ materials	17 3%	1–10	1	–	(i)
As above 0.20	1	1	As above	2.94	5#	As above	18 3%	1–10	1	–	(ii)
As above 0.20	1	1	As above	2.94	5#	As above	20 3%	1–10	2	–	(iii)
As above 0.20	1	1	As above	2.94	5#	As above	20 3%	1–10	1	–	(iv)
As above 0.20	1	1	As above	2.94	5#	As above	20 3%	1–10	1	–	(v)
As above 0.20	1	1	As above	2.94	5#	As above	22 3%	1–10	2	–	(vi)
											b
As above 0.20	1	1	As above	2.94	5	As above	19 3%	1–10	6	–	(i)
As above 0.20	1	1	As above	3.05	5	As above	21 3%	1–10	9	–	(ii)
As above 0.20	1	1	As above	3.45	5	As above	28 3%	1–10	12	–	(iii)
As above 0.20	1	1	As above	3.05	5	As above	19 3%	1–10	11	–	(iv)
As above 0.20	1	1	As above	3.45	5	As above	21 3%	1–10	15	–	(v)
											c
As above 0.20	1	1	–	–	–	As above	10 3%	1–10	1	–	(i)
As above 0.20	1	1	–	–	–	As above	11 3%	1–10	1	–	(ii)

Commencing year 15

No	2.C.3 Roof drainage — *continued* Description		Capital	Cleaning	Cyclical	Repairs	Total
2	GUTTERS — *continued*	m					
c	UPVC, BS 4576, brackets screwed to timber — *continued*						
	(iii) 152 half round		10.42	3.23	–	0.33	13.98
	(iv) 114 rectangular		6.69	3.23	–	0.26	10.18
d	Galvanised pressed steel, BS 1091, 3 thick, bolted joints; screwed to timber						
	(i) 200 × 100 box gutter		25.54	3.23	13.19	0.68	42.64
	(ii) 380 boundary wall gutter		26.70	3.23	18.47	0.71	49.11
	(iii) 460 boundary wall gutter		32.86	3.23	21.33	0.86	58.28
3	ROOF OUTLETS	Nr					
a	Aluminium, circular, flat or domed, jointed to pipe						
	(i) 75 pipe		44.40	80.81	–	1.10	126.31
	(ii) 100 pipe		58.19	80.81	–	1.39	140.39
b	Aluminium, D shape balcony outlet, flat or domed, jointed to pipe						
	(i) 75 pipe		47.00	80.81	–	1.17	128.98
	(ii) 100 pipe		56.77	80.81	–	1.39	138.97
c	Cast iron, circular, flat, jointed to pipe						
	(i) 75 pipe		46.54	80.81	–	1.15	128.50
	(ii) 100 pipe		54.70	80.81	–	1.35	136.86

Cleaning			Cyclical			Repairs					
Annual Cost	*Frequency No*	*Yr*	*Work*	*Cost*	*Cycle in Years*	*Cause/ Action*	*Cost (£) and Risk*	*Occurs in Years*	*kg/m*	*R value*	*No*
											2
											c
Clean out and wash 0.20	1	1	–	–	–	Defective work/ materials	15 3%	1–10	1	–	(iii)
As above 0.20	1	1	–	–	–	As above	12 3%	1–10	1	–	(iv)
											d
As above 0.20	1	1	Repaint	4.65	5	As above	31 3%	1–10	10	–	(i)
As above 0.20	1	1	As above	6.51	5	As above	32 3%	1–10	19	–	(ii)
As above 0.20	1	1	As above	7.52	5	As above	39 3%	1–10	22	–	(iii)
											3
											a
As above 5.00	2	1	–	–	–	Replace gratings, repair joints	50 3%	1–10	–	–	(i)
As above 5.00	2	1	–	–	–	As above	63 3%	1–10	–	–	(ii)
											b
As above 5.00	2	1	–	–	–	As above	53 3%	1–10	–	–	(i)
As above 5.00	2	1	–	–	–	As above	63 3%	1–10	–	–	(ii)
											c
As above 5.00	2	1	–	–	–	As above	52 3%	1–10	–	–	(i)
As above 5.00	2	1	–	–	–	As above	61 3%	1–10	–	–	(ii)

2.C.4 Roof lights

No	Description		Capital	Cleaning	Cyclical	Repairs	Total
1	CONCRETE	m^2					
a	Standard reinforced precast flat unit; clear glass lens 198 × 198 at 254 centres, bedded; pointed						
	(i) 125 thick, single lens, 2.6 – 3.4m span		309.13	6.95	–	0.63	316.71
	(ii) 150 thick, double lens, 3.2 – 4.0m span		419.17	6.95	–	0.85	426.97
2	TIMBER	Nr					
a	Standard sloping roof window; treated pine frame and opening light; aluminium trim; ironmongery; double glazing; flashings and soakers						
	(i) 550 × 700 ($0.39m^2$)		156.85	15.84	7.80	0.94	181.43
	(ii) 780 × 980 ($0.76m^2$)		194.82	31.35	10.92	1.14	238.23
	(iii) 1340 × 1400 ($1.88m^2$)		292.46	76.93	17.16	1.70	388.25
3	ALUMINIUM	Nr					
a	Standard sloping roof window; frame and opening light, integral internal lining, integral flashings and soakers; ironmongery; double glazing						
	(i) 540 × 1050 ($0.57m^2$)		170.67	23.27	–	0.51	194.45
	(ii) 710 × 1050 ($0.75m^2$)		205.05	30.54	–	0.60	236.19
	(iii) 940 × 1050 ($0.99m^2$)		241.51	40.56	–	0.70	282.77

Cleaning			*Cyclical*			*Repairs*					
Annual Cost	*Frequency No*	*Yr*	*Work*	*Cost*	*Cycle in Years*	*Cause/ Action*	*Cost (£) and Risk*	*Occurs in Years*	*kg/m²*	*R value*	*No*
											1
											a
Clean glass both sides 0.43	2	1	–	–	–	Defective work	340 0.25%	1–10	186	0.21	(i)
As above 0.43	2	1	–	–	–	As above	461 0.25%	1–10	239	0.43	(ii)
									kg		2
											a
As above 0.98	12	1	Re-decorate	2.75	5	Defective work, rot, woodworm	174 2%	1–60	47	0.44	(i)
As above 1.94	12	1	As above	3.85	5	As above	211 2%	1–60	38	0.40	(ii)
As above 4.76	12	1	As above	6.05	5	As above	315 2%	1–60	30	0.37	(iii)
									kg		3
											a
As above 1.44	12	1	–	–	–	Defective work	188 1%	1–60	40	0.42	(i)
As above 1.89	12	1	–	–	–	As above	222 1%	1–60	30	0.40	(ii)
As above 2.51	12	1	–	–	–	As above	259 1%	1–60	28	0.40	(iii)

No	Description	Capital	Cleaning	Cyclical	Repairs	Total
	2.C.4 Roof lights — *continued*					
4	UPVC Nr					
a	Standard square or rectangular dome; timber kerb; plugged and screwed; plywood lining painted; roof skirting/flashing					
	(i) 900 × 600 (0.54m²) single skin	123.52	3.72	–	0.36	127.60
	(ii) 1200 × 900 (1.08m²) single skin	159.74	7.43	–	0.47	167.64
	(iii) 1800 × 1200 (2.16m²) single skin	255.97	14.71	–	0.74	271.42
	(iv) 900 × 600 (0.54m²) double skin; aluminium upstand with two ventilators	351.37	3.72	–	0.98	356.07
	(v) 1200 × 900 (1.08m²) double skin; aluminium upstand with two ventilators	422.03	7.43	–	1.18	430.64
	(vi) 1800 × 1200 (2.16m²) double skin; aluminium upstand with two ventilators	645.25	14.71	–	1.79	661.75
	2.D Stairs					
1	REINFORCED CONCRETE Nr					
a	Two flights with half space landing; one storey high, 2700 rise; 1000 wide; 21N in situ concrete, reinforcement; formwork; granolithic finish; plaster and painted solfits; mild steel painted balustrade one side	1750.00	3636.23	59.80	0.54	5446.57
2	SOFTWOOD Nr					
a	Two flights with half space landing; one storey high, 2700 rise; 900 wide; plasterboard, plaster and painted soffits; painted balustrade one side	950.00	3232.20	137.22	3.08	4322.50

Cleaning			Cyclical			Repairs					
Annual Cost	Frequency No	Yr	Work	Cost	Cycle in Years	Cause/ Action	Cost (£) and Risk	Occurs in Years	kg	R value	No
											4
											a
Clean glass both sides 0.23	2	1	–	–	–	Defective work	135 1%	1–60	30	0.18	(i)
As above 0.46	2	1	–	–	–	As above	176 1%	1–60	25	0.18	(ii)
As above 0.91	2	1	–	–	–	As above	276 1%	1–60	20	0.18	(iii)
As above 0.23	2	1	–	–	–	As above	364 1%	1–60	45	0.36	(iv)
As above 0.46	2	1	–	–	–	As above	438 1%	1–60	40	0.36	(v)
As above 0.91	2	1	–	–	–	As above	666 1%	1–60	35	0.36	(vi)
											1
Dust/ sweep 170 Wash 55	250 50	1 1	Repaint	50.00	10	Defective work/ materials	1950 0.15%	10–60	6500 kg total	–	a
											2
As above 150 50	250 50	1 1	As above	60.00	6	As above Dry rot, woodworm	1050 0.15% 1050 1%	10–60 10–40	300 kg total	–	a

2 SUPERSTRUCTURE

2.E.A External wall structure

No	Description	Capital	Cleaning	Cyclical	Repairs	Total
	NOTE. Add on external wall cladding and internal wall finishings as required					
1	WALLING m^2					
a	Reinforced concrete in situ P25N; hy reinforcement 13 kg/m^2; formwork basic					
	(i) 100 thick	58.19	–	–	0.02	58.21
	(ii) 140 thick	65.55	–	–	0.02	65.57
	(iii) 240 thick	83.96	–	–	0.03	83.99
b	Concrete blocks, dense aggregate; gauged mortar (1:2:9)					
	(i) 100 thick, solid	18.12	–	–	0.01	18.13
	(ii) 140 thick, solid	23.83	–	–	0.01	23.84
	(iii) 215 thick, hollow	31.19	–	–	0.02	31.21
c	Concrete blocks, lightweight aerated; gauged mortar (1:2:9)					
	(i) 100 thick	16.61	–	–	0.01	16.62
	(ii) 140 thick	21.36	–	–	0.01	21.37
	(iii) 215 thick	30.51	–	–	0.02	30.53
d	Concrete blocks, lightweight aerated; smooth faced; gauged mortar (1:2:9); flush pointed					
	(i) 100 thick	21.26	3.23	–	0.01	24.50
	(ii) 140 thick	27.28	3.23	–	0.01	30.52
	(iii) 215 thick	38.54	3.23	–	0.02	41.79

Cleaning			Cyclical			Repairs					
Annual Cost	Frequency No	Yr	Work	Cost	Cycle in Years	Cause/ Action	Cost (£) and Risk	Occurs in Years	kg/m²	R value	No
											1
											a
–	–	–	–	–	–	Defective work/ materials	63 0.15%	10–60	245	0.07	(i)
–	–	–	–	–	–	As above	72 0.15%	10–60	343	0.10	(ii)
–	–	–	–	–	–	As above	95 0.15%	10–60	588	0.17	(iii)
											b
–	–	–	–	–	–	As above	22 0.25%	10–60	230	0.06	(i)
–	–	–	–	–	–	As above	29 0.25%	10–60	322	0.09	(ii)
–	–	–	–	–	–	As above	37 0.25%	10–60	295	0.26	(iii)
											c
–	–	–	–	–	–	As above	20 0.25%	10–60	60	0.63	(i)
–	–	–	–	–	–	As above	26 0.25%	10–60	84	0.88	(ii)
–	–	–	–	–	–	As above	37 0.25%	10–60	129	1.34	(iii)
											d
Dust internal 0.20	6	1	–	–	–	As above	24 0.25%	10–60	60	0.63	(i)
As above 0.20	6	1	–	–	–	As above	32 0.25%	10–60	84	0.88	(ii)
As above 0.20	6	1	–	–	–	As above	46 0.25%	10–60	129	1.34	(iii)

2.E.A External wall structure – *continued*

No	Description	Capital	Cleaning	Cyclical	Repairs	Total
1	WALLING – *continued* m²					
e	Bricks, common; gauged mortar (1:1:6)					
	(i) half brick wall	23.98	–	–	0.01	23.99
	(ii) one brick wall	44.51	–	–	0.02	44.53
	(iii) one and a half brick wall	63.21	–	–	0.03	63.24
f	Bricks, facings, sand faced (PC £120.00/K); pointed; gauged mortar (1:1:6)					
	(i) half brick wall, faced one side, stretcher bond	29.13	3.23	–	0.02	32.38
	(ii) half brick wall, faced one side, flemish bond, snapped headers	32.71	3.23	–	0.02	35.96
g	Bricks, facings, machine made (PC £300.00/K); pointed, gauged mortar (1:1:6)					
	(i) half brick wall, faced one side, stretcher bond	43.07	3.23	–	0.02	46.32
	(ii) half brick wall, faced one side, flemish bond, snapped headers	47.54	3.23	–	0.02	50.79
h	Bricks, facings, hand made (PC £490.00/K); pointed; gauged mortar (1:1:9)					
	(i) half brick wall, faced one side, stretcher bond	57.79	3.23	–	0.03	61.05
	(ii) half brick wall, faced one side, flemish bond, snapped headers	62.77	3.23	–	0.03	66.03
j	Glass blocks, hollow, cross reeded, clear or similar; cement and sand joints; pointed both sides in mastic					
	(i) 190 × 190 × 80	193.04	6.87	–	0.09	200.00
	(ii) 240 × 240 × 80; 6mm stainless steel rods in joints	261.99	6.87	–	0.12	268.98

Cleaning			Cyclical			Repairs					
Annual Cost	Frequency No	Yr	Work	Cost	Cycle in Years	Cause/ Action	Cost (£) and Risk	Occurs in Years	kg/m^2	R value	No
											1
											e
–	–	–	–	–	–	Defective work/ materials	29 0.25%	10–60	192	0.12	(i)
–	–	–	–	–	–	As above	54 0.25%	10–60	392	0.26	(ii)
–	–	–	–	–	–	As above	76 0.25%	10–60	583	0.39	(iii)
											f
Dust internal 0.20	6	1	–	–	–	As above	33 0.25%	10–60	192	0.13	(i)
As above 0.20	6	1	–	–	–	As above	37 0.25%	10–60	192	0.13	(ii)
											g
As above 0.20	6	1	–	–	–	As above	47 0.25%	10–60	192	0.13	(i)
As above 0.20	6	1	–	–	–	As above	52 0.25%	10–60	192	0.13	(ii)
											h
As above 0.20	6	1	–	–	–	As above	62 0.25%	10–60	192	0.12	(i)
As above 0.20	6	1	–	–	–	As above	67 0.25%	10–60	192	0.12	(ii)
											j
Dust both sides 0.40	6	1	–	–	–	Minor defects	199 0.25%	10–60	67	0.12	(i)
Wash both sides 0.18	1	6									
As above 0.40	6	1	–	–	–	As above	267 0.25%	10–60	68	0.12	(ii)
0.18	1	6									

2.E.A External wall structure – *continued*

No	Description		Capital	Cleaning	Cyclical	Repairs	Total
1	WALLING – *continued*	m²					
k	Natural limestone, Cotswold, random rubble; gauged mortar (1:2:9), pointed one side						
	(i) 275 thick, uncoursed		58.77	–	–	0.03	58.80
	(ii) 350 thick, uncoursed		71.24	–	–	0.04	71.28
	(iii) 114 thick, coursed		66.13	–	–	0.03	66.16
	(iv) 150 thick, coursed		74.07	–	–	0.04	74.11
l	Natural limestone, Portland, ashlar; gauged mortar (1:2:9); bedded against other work; cramps; stone dust pointed one side						
	(i) 50 thick		192.43	0.04	–	0.09	192.56
	(ii) 75 thick		263.62	0.04	–	0.12	263.78
m	Reconstructed limestone blocks, Cotswold, ashlar or rough hewn; coloured gauged mortar (1:2:9), pointed one side, 100 thick						
	(i) random, uncoursed		33.84	0.04	–	0.02	33.90
	(ii) squared, coursed		36.04	0.04	–	0.02	36.10
2	CAVITIES	m²					
a	Open cavity, 50						
	(i) polyproplyene ties		0.76	–	–	0.01	0.77
	(ii) galvanised butterfly ties		1.14	–	–	0.01	1.15
	(iii) stainless steel twisted ties		1.82	–	–	0.01	1.83

Cleaning			Cyclical			Repairs					
Annual Cost	Frequency No	Yr	Work	Cost	Cycle in Years	Cause/ Action	Cost (£) and Risk	Occurs in Years	kg/m²	R value	No
											k
–	–	–	–	–	–	Minor defects	69 0.25%	10–60	600	0.18	(i)
–	–	–	–	–	–	As above	84 0.25%	10–60	763	0.23	(ii)
–	–	–	–	–	–	As above	69 0.25%	10–60	250	0.08	(iii)
–	–	–	–	–	–	As above	79 0.25%	10–60	330	0.10	(iv)
											l
Specialist clean external 0.25	1	30	–	–	–	As above	195 0.25%	10–60	121	0.03	(i)
As above 0.25	1	30	–	–	–	As above	266 0.25%	10–60	177	0.05	(ii)
											m
As above 0.25	1	30	–	–	–	As above	37 0.25%	10–60	224	0.07	(i)
As above 0.25	1	30	–	–	–	As above	40 0.25%	10–60	224	0.07	(ii)
											2
											a
–	–	–	–	–	–	Tie failure	10 0.25%	10–60	–	0.18	(i)
–	–	–	–	–	–	As above	10 0.25%	10–60	–	0.18	(ii)
–	–	–	–	–	–	As above	10 0.25%	10–60	–	0.18	(iii)

2.E.A External wall structure — *continued*

No	Description		Capital	Cleaning	Cyclical	Repairs	Total
3	CAVITY INSULATION	m²					
a	Glassfibre, flanged, pinned to timber						
	(i) 60 thick		3.20	–	–	0.01	3.21
	(ii) 80 thick		3.77	–	–	0.01	3.78
	(iii) 100 thick		4.29	–	–	0.01	4.30
b	Polystyrene expanded board, fixed with adhesive or ties						
	(i) 25 thick		3.64	–	–	0.01	3.65
	(ii) 50 thick		5.36	–	–	0.01	5.37
c	Mineral wool filling to cavity						
	(i) 50 thick		3.74	–	–	0.01	3.75
	(ii) 75 Thick		5.61	–	–	0.01	5.62
d	Vermiculite bead filling to cavity						
	(i) 50 thick		4.00	–	–	0.01	4.01
	(ii) 75 thick		6.01	–	–	0.01	6.02

Cleaning			Cyclical			Repairs					
Annual Cost	*Frequency No*	*Yr*	*Work*	*Cost*	*Cycle in Years*	*Cause/ Action*	*Cost (£) and Risk*	*Occurs in Years*	*kg/m²*	*R value*	*No*
											3
											a
–	–	–	–	–	–	Failure	23 0.25%	10–60	2	1.50	(i)
–	–	–	–	–	–	As above	24 0.25%	10–60	2	2.00	(ii)
–	–	–	–	–	–	As above	24 0.25%	10–60	3	2.50	(iii)
											b
–	–	–	–	–	–	As above	23 0.25%	10–60	1	0.68	(i)
–	–	–	–	–	–	As above	25 0.25%	10–60	2	1.39	(ii)
											c
–	–	–	–	–	–	As above	24 0.25%	10–60	3	1.39	(i)
–	–	–	–	–	–	As above	25 0.25%	10–60	4	2.08	(ii)
											d
–	–	–	–	–	–	As above	24 0.25%	10–60	5	0.77	(i)
–	–	–	–	–	–	As above	26 0.25%	10–60	10	1.59	(ii)

2.E.B External wall coverings

No	Description	Capital	Cleaning	Cyclical	Repairs	Total
1	CAST FINISHES m^2					
a	Extra over in situ concrete for superfine finish	4.65	–	–	–	4.65
b	Extra over in situ concrete for exposed aggregate face by mechanical means	7.84	–	–	–	7.84
c	Extra over in situ concrete for formed feature finish	6.20	–	–	–	6.20
2	APPLIED FINISHES m^2					
a	Slate slabs, natural blue grey, 450 × 600 × 30, fine rubbed finish, bedded in cement and sand	171.60	–	–	0.38	171.98
b	Natural stone, Bath or York, 75 thick, stainless steel fixing cramps	202.40	0.04	–	0.45	202.89
c	Mosaic tiling, simple patterns, glass or ceramic, fixed with adhesive; levelling backing	64.35	0.28	–	0.15	64.78
d	Cement and sand render, 15 thick, plain finish	7.32	–	5.84	0.02	13.18
e	Cement and sand, pebbledash finish	10.15	–	8.11	0.03	18.29
f	Tyrolean finish, 3 coats; cement and sand backing	15.51	–	6.78	0.04	22.33
g	Mineralite finish; cement and sand backing	33.00	–	13.08	0.08	46.16
h	Specialist waterproof rendering, 3 coat, 20 thick, plain face	27.80	–	5.84	0.07	33.71
j	Textured masonry paint, two coats	4.18	–	16.12	–	20.30

Cleaning			Cyclical			Repairs					
Annual Cost	*Frequency No*	*Yr*	*Work*	*Cost*	*Cycle in Years*	*Cause/ Action*	*Cost (£) and Risk*	*Occurs in Years*	*kg/m²*	*R value*	*No*
											1
–	–	–	–	–	–	–	–	–	–	–	a
–	–	–	–	–	–	–	–	–	–	–	b
–	–	–	–	–	–	–	–	–	–	–	c
											2
–	–	–	–	–	–	Loss of adhesion, frost action	176 1%	10–50	113	0.05	a
Jet scrub 0.25	1	30	–	–	–	Loose fixings, frost action	206 1%	10–50	165	0.05	b
Washing 0.10	1	5	–	–	–	Loss of adhesion	69 1%	10–50	41	0.02	c
–	–	–	Repair cracks	2.06	5	As above	10 1%	10–50	26	0.03	d
–	–	–	Repair cracks, replace stones	2.86	5	As above	12 1%	10–50	59	0.07	e
–	–	–	Repair cracks, replace patches	2.39	5	As above	19 1%	10–50	32	0.05	f
–	–	–	As above	4.61	5	As above	36 1%	10–50	27	0.04	g
–	–	–	Repair cracks	2.06	5	As above	31 1%	10–50	32	0.04	h
–	–	–	Repaint	4.40	4	–	–	–	3	–	j

2.E.B External wall coverings – *continued*

No	Description		Capital	Cleaning	Cyclical	Repairs	Total
3	BOARDING	m²					
a	Wrot softwood, shiplap, 25 thick; sawn softwood battens, 600 centres; vapour barrier; oil paint 3 coats		26.31	–	6.98	0.08	33.37
b	Western red cedar, shiplap, 25 thick; sawn softwood battens, 600 centres; vapour barrier; preservative 2 coats		34.46	–	15.79	0.10	50.35
c	Plastic weatherboarding, insulated, clips; sawn softwood battens, 450 centres; vapour barrier		33.26	0.28	–	0.78	34.32
4	FLAT SHEETING	m²					
a	Zinc, 14 gauge, 0.81 thick, seamed joints, clips; 12 exterior ply; underlay; additional bearers		66.99	–	–	0.08	67.07
b	Lead, Code 3, 1.32 thick; 18 exterior ply, composite; seamed joints; additional bearers		101.01	–	–	0.12	101.13
c	Zinc, 12 gauge, 0.65 thick; 18 exterior ply, composite; seamed joints; additional bearers		95.51	–	–	0.11	95.62
5	PROFILED SHEETING	m²					
a	Fibre cement corrugated, natural colour; additional bearers; to timber		14.12	–	1.34	0.08	15.54
b	Fibre cement corrugated, natural colour; additional bearers; to steel		18.11	–	1.44	0.08	19.63
c	Fibre cement corrugated, acrylic coated; additional bearers; to timber		15.59	0.23	1.68	0.10	17.60
d	Fibre cement corrugated, acrylic coated; additional bearers; to steel		19.45	0.23	1.53	0.09	21.30
e	Fibre cement corrugated, natural colour, bonded to insulation 60 thick, sandwich; additional bearers; to steel		31.55	–	2.33	0.16	34.04

Cleaning			Cyclical			Repairs					
Annual Cost	*Frequency No*	*Yr*	*Work*	*Cost*	*Cycle in Years*	*Cause/ Action*	*Cost (£) and Risk*	*Occurs in Years*	*kg/m²*	*R value*	*No*
											3
–	–	–	Repaint	2.46	5	Dry rot/wet rot/ woodworm	29 1%	10–40	20	0.37	a
–	–	–	Re-treat	2.02	2	As above	37 1%	10–40	16	0.42	b
Wash 0.10	1	5	–	–	–	Physical damage, surface deterioration	36 8%	1–60	7	0.80	c
											4
–	–	–	–	–	–	Defective work	69 0.2%	1–20	16	0.10	a
–	–	–	–	–	–	As above	103 0.2%	1–20	29	0.14	b
–	–	–	–	–	–	As above	98 0.2%	1–20	18	0.14	c
											5
–	–	–	Renew	13.84	40	Physical damage	15 2%	1–60	12	0.03	a
–	–	–	As above	14.89	40	As above	15 2%	1–60	15	0.03	b
Wash 0.08	1	5	As above	17.36	40	As above	18 2%	1–60	13	0.04	c
As above 0.08	1	5	As above	15.74	40	As above	17 2%	1–60	16	0.04	d
–	–	–	As above	23.97	40	As above	29 2%	1–60	25	2.45	e

2.E.B External wall coverings – *continued*

No	*Description*		*Capital*	*Cleaning*	*Cyclical*	*Repairs*	*Total*
5	PROFILED SHEETING – *continued*	m²					
f	Steel, PVC colour coated, average profile, bolted laps; additional bearers; to timber		17.80	0.23	–	0.08	18.11
g	Steel, PVC colour coated, average profile, bolted laps; additional bearers; to steel		21.35	0.23	–	0.07	21.65
h	Aluminium anodised or pre-coated, average profile; additional bearers; to timber		21.30	0.23	–	0.09	21.62
j	Aluminium anodised or pre-coated, average profile; additional bearers; to steel		24.78	0.23	–	0.09	25.10
k	G.R.P. panels, plain finish/imitation aggregate finish; to timber		118.16	0.28	–	2.89	121.33
l	G.R.P. panels, decorative finish, 60 insulation filled; aluminium fixings; vapour barrier; additional bearers; to steel		125.06	0.28	–	2.63	127.97
m	Precast concrete panels, decorative surface, 125 thick; stainless steel fixings; additional bearers; to steel		323.60	–	4.19	0.77	328.56
6	HANGINGS	m²					
a	Clay plain tiles, machine made; 65 lap; softwood battens; felt underlay		36.49	–	–	1.05	37.54
b	Clay interlocking tiles, 100 lap; softwood battens; felt underlay		26.84	–	–	0.78	27.62
c	Concrete plain tiles, 65 lap; softwood battens; felt underlay		31.48	–	–	0.92	32.40

Cleaning			Cyclical			Repairs					
Annual Cost	Frequency No	Yr	Work	Cost	Cycle in Years	Cause/ Action	Cost (£) and Risk	Occurs in Years	kg/m²	R value	No
											5
Wash 0.08	1	5	–	–	–	Physical damage	20 1.5%	1–60	8	ignore	f
As above 0.08	1	5	–	–	–	As above	18 1.5%	1–60	11	ignore	g
As above 0.08	1	5	–	–	–	As above	23 1.5%	1–60	5	ignore	h
As above 0.08	1	5	–	–	–	As above	22 1.5%	1–60	8	ignore	j
As above 0.10	1	5	–	–	–	Physical damage, surface deterioration	134 8%	1–60	5	0.03	k
As above 0.10	1	5	–	–	–	As above	122 8%	1–60	6	2.30	l
–	–	–	Renew jointing	24.07	30	Defective fixings or joints	352 1%	10–50	293	0.09	m
											6
–	–	–	–	–	–	Defective materials, wind damage	39 10%	1–60	72	0.24	a
–	–	–	–	–	–	As above	29 10%	1–60	64	0.23	b
–	–	–	–	–	–	As above	34 10%	1–60	101	0.24	c

2.E.B External wall coverings – *continued*

No	Description		Capital	Cleaning	Cyclical	Repairs	Total
6	HANGINGS – *continued*	m²					
d	Concrete interlocking tiles, 75 lap; softwood battens; felt underlay		15.85	–	–	0.48	16.33
e	Natural slates, Welsh blue, 75 lap; softwood battens; felt underlay		38.86	–	–	1.10	39.96
f	Slates reconstituted, interlocking, 50 lap; softwood battens; felt underlay		26.38	–	–	0.78	27.16
g	Slabs reconstituted, stone, Cotswold/Moordale, 75 lap; softwood battens; felt underlay		44.92	–	–	1.29	46.21
h	Slates; fibre cement; 75 lap; softwood battens; felt underlay		28.75	–	–	0.83	29.58
j	Shingles, Red Cedar, treated; 75 lap; softwood battens; felt underlay		41.05	–	13.53	1.39	55.97
7	CURTAIN WALLING	m²					
a	Steel framework, galvanised, colour coated (excluding infill)		178.20	–	–	–	178.20
b	Aluminium framework, anodised (excluding infill)		212.30	–	–	–	212.30
c	Infill – 6mm float single glazing – some obscure panels		28.60	6.79	9.84	0.58	45.81
d	Infill – 6mm float double glazing – some obscure panels		77.44	6.79	9.84	1.37	95.44
e	Infill – 6mm float/6mm anti-sun double glazing – some obscure panels		121.00	6.79	9.84	2.08	139.71

Cleaning			Cyclical			Repairs					
Annual Cost	*Frequency No*	*Yr*	*Work*	*Cost*	*Cycle in Years*	*Cause/ Action*	*Cost (£) and Risk*	*Occurs in Years*	*kg/m²*	*R value*	*No*
											6
–	–	–	–	–	–	Defective materials, wind damage	18 10%	1–60	79	0.18	d
–	–	–	–	–	–	As above	41 10%	1–60	69	0.21	e
–	–	–	–	–	–	As above	29 10%	1–60	73	0.21	f
–	–	–	–	–	–	Defective materials, frost damage	48 10%	1–60	122	0.24	g
–	–	–	–	–	–	Defective materials, wind damage	31 10%	1–60	18	0.48	h
–	–	–	Re-treat Renew	2.02 43.34	3 44	Loosening, splitting, wind damage	43 12%	1–60	26	0.50	j
											7
See infill			See infill			See infill			16	Ignore	a
See infill			See infill			See infill			7	Ignore	b
Wash both sides 0.42	2	1	Replace gaskets	24.07	20	Physical damage, faulty joints	36 6%	1–60	15	0.01	c
As above 0.42	2	1	As above	24.07	20	As above	85 6%	1–60	31	0.18	d
As above 0.42	2	1	As above	24.07	20	As above	129 6%	1–60	31	0.18	e

2.E.B External wall coverings – *continued*

No	*Description*		*Capital*	*Cleaning*	*Cyclical*	*Repairs*	*Total*
8	ADDITIONAL INSULATION	m^2					
a	Glassfibre/mineral wool mat 60, supported by other work		3.63	–	–	–	3.63
b	Glassfibre/mineral wool mat 100, supported by other work		4.66	–	–	–	4.66
c	Polystyrene board 50, fixed with adhesive		6.28	–	–	–	6.28

Cleaning			Cyclical			Repairs					
Annual Cost	Frequency No	Yr	Work	Cost	Cycle in Years	Cause/ Action	Cost (£) and Risk	Occurs in Years	kg/m^2	R value	No
											8
–	–	–	–	–	–	–	–	–	3	1.82	a
–	–	–	–	–	–	–	–	–	5	3.03	b
–	–	–	–	–	–	–	–	–	2	1.52	c

2.F.1 Windows

No	Description	Capital	Cleaning	Cyclical	Repairs	Total
1	SOFTWOOD m^2					
a	Treated softwood casements, primed and painted three coats; fixing cramps; average 1.44m^2; 4mm glazed					
	(i) standard, BS 644, fixed lights	69.40	44.28	30.23	0.49	144.40
	(ii) standard, BS 644, side hung opening lights; latch and stay	80.49	44.28	47.00	0.55	172.32
	(iii) standard, BS 644, top hung opening lights; stay	82.12	44.28	46.53	0.57	173.50
	(iv) non-standard, BS 644, opening lights; divided into medium panes; latch and stay	157.60	44.28	47.00	0.98	249.86
	(v) standard, high performance, opening pivot lights; stay	124.49	44.28	47.79	0.79	217.35
	(vi) non-standard; opening pivot lights; stay	200.17	44.28	47.79	1.20	293.44
b	Treated softwood, double hung sash window; primed and painted three coats; fixing cramps; average 1.44m^2; 4mm glazed; catch					
	(i) standard, high performance cased frame, spiral balancing	171.24	44.28	35.75	1.05	252.32
	(ii) purpose made, solid frame, spiral balancing	213.08	44.28	35.75	1.27	294.38
	(iii) purpose made, boxed frame, cords and weights	185.75	44.28	36.00	1.12	267.15

Cleaning		*Cyclical*			*Repairs*					
Annual Cost	*Frequency No Yr*	*Work*	*Cost*	*Cycle in Years*	*Cause/ Action*	*Cost (£) and Risk*	*Occurs in Years*	*kg/m²*	*R value*	*No*
										1
										a
Clean both sides 2.74	12 1	Overhaul Paint int. Paint ext.	5.10 5.50 5.91	30 6 5	Replace frame	91 2%	1–60	25	0.19*	(i)
As above 2.74	12 1	As above	28.36 6.61 9.50	30 6 5	Replace frame and/or ironmongery	102 2%	1–60	32	0.20*	(ii)
As above 2.74	12 1	As above	25.64 6.61 9.50	30 6 5	As above	105 2%	1–60	32	0.20*	(iii)
As above 2.74	12 1	As above	28.36 6.61 9.50	30 6 5	As above	182 2%	1–60	32	0.20*	(iv)
As above 2.74	12 1	As above	32.87 6.61 9.50	30 6 5	As above	146 2%	1–60	32	0.20*	(v)
As above 2.74	12 1	As above	32.87 6.61 9.50	30 6 5	As above	222 2%	1–60	32	0.20*	(vi)
										b
As above 2.74	12 1	As above	36.95 5.49 5.91	30 6 5	As above	194 2%	1–60	46	0.20*	(i)
As above 2.74	12 1	As above	36.95 5.49 5.91	30 6 5	As above	235 2%	1–60	46	0.20*	(ii)
As above 2.74	12 1	As above	38.37 5.49 5.91	30 6 5	As above	208 2%	1–60	66	0.20*	(iii)

*Including surface resistance.

2.F.1 Windows – *continued*

No	*Description*		*Capital*	*Cleaning*	*Cyclical*	*Repairs*	*Total*
2	HARDWOOD	m^2					
a	Phillipine mahogany or similar casements; stain preservative finish; fixing cramps; average 1.44m^2; 4mm glazed						
	(i) standard, fixed lights		87.22	44.28	19.17	0.59	151.26
	(ii) standard, side hung, fastener, stay		131.13	44.28	31.01	0.82	207.24
	(iii) standard, top hung, stay		117.37	44.28	30.38	0.75	192.78
	(iv) non-standard, opening lights, divided into medium panes, latch and stay		213.33	44.28	31.01	1.27	289.89
b	West African mahogany or similar double hung sash window; stain preservative finish; fixing cramps; average 1.44m^2: 4mm glazed; catch						
	(i) purpose made; solid frame; spiral balancing		290.60	44.28	25.79	1.68	362.36
	(ii) purpose made; boxed frame; chains and weights		323.71	44.28	25.36	1.86	395.21

Cleaning			Cyclical			Repairs					
Annual Cost	*Frequency No*	*Yr*	*Work*	*Cost*	*Cycle in Years*	*Cause/ Action*	*Cost (£) and Risk*	*Occurs in Years*	*kg/m²*	*R value*	*No*
											2
											a
Clean both sides 2.74	12	1	Overhaul Varnish int. Treat ext.	6.35 4.13 3.04	30 6 5	Replace frame	109 2%	1–60	36	0.19*	(i)
As above 2.74	12	1	As above	33.10 4.95 4.91	30 6 5	Replace frame and/or ironmongery	153 2%	1–60	45	0.20*	(ii)
As above 2.74	12	1	As above	29.48 4.95 4.91	30 6 5	As above	140 2%	1–60	45	0.20*	(iii)
As above 2.74	12	1	As above	33.10 4.95 4.91	30 6 5	As above	235 2%	1–60	45	0.20*	(iv)
											b
As above 2.74	12	1	As above	44.37 4.13 3.04	30 6 5	As above	312 2%	1–60	54	0.20*	(i)
As above 2.74	12	1	As above	41.92 4.13 3.04	30 6 5	As above	345 2%	1–60	80	0.20*	(ii)

*Including surface resistance

2.F.1 Windows – *continued*

No	*Description*		*Capital*	*Cleaning*	*Cyclical*	*Repairs*	*Total*
3	STEEL	m^2					
a	Standard, galvanised; average 1.32m^2; cill; fixing lugs; 4mm glazed; primed and painted three coats						
	(i) fixed light		81.27	44.28	25.87	0.26	151.68
	(ii) side/top hung lights; catch and stay		146.66	44.28	42.68	0.44	234.06
	(iii) composite fixed/opening; catch and stay		127.39	44.28	42.68	0.39	214.74
	(iv) reversible; catch and stay		151.81	44.28	42.68	0.45	239.22
b	Standard, galvanised; average 1.32m^2; cill; fixing lugs; 4mm glazed; powder polyester coated at works						
	(i) fixed light		75.96	44.28	11.66	0.25	132.15
	(ii) side/top hung lights; catch and stay		152.09	44.28	23.02	0.46	219.85
	(iii) composite fixed/opening; catch and stay		131.09	44.28	23.02	0.40	198.79
	(iv) reversible; catch and stay		155.45	44.28	23.02	0.47	223.22

Cleaning			Cyclical			Repairs					
Annual Cost	*Frequency No*	*Yr*	*Work*	*Cost*	*Cycle in Years*	*Cause/ Action*	*Cost (£) and Risk*	*Occurs in Years*	*kg/m²*	*R value*	*No*
											3
											a
Clean both sides 2.74	12	1	Paint int. Paint ext.	4.85 5.21	6 5	Replace frame	98 1%	1–60	26	0.18*	(i)
As above 2.74	12	1	Overhaul Paint int. Paint ext.	21.77 4.85 9.80	30 6 5	Replace frame and/or ironmongery	164 1%	1–60	42	0.18*	(ii)
As above 2.74	12	1	As above	21.77 4.85 9.80	30 6 5	As above	144 1%	1–60	37	0.18*	(iii)
As above 2.74	12	1	As above	21.77 4.85 9.80	30 6 5	As above	168 1%	1–60	42	0.18*	(iv)
											b
As above 2.74	12	1	Clean int. Paint ext.	1.85 4.85	6 5#	Replace frame	92 1%	1–60	26	0.18*	(i)
As above 2.74	12	1	Overhaul Clean int. Paint ext.	21.77 1.85 9.80	30 6 5#	Replace frame and/or ironmongery	169 1%	1–60	42	0.18*	(ii)
As above 2.74	12	1	As above	21.77 1.85 9.80	30 6 5#	As above	149 1%	1–60	37	0.18*	(iii)
As above 2.74	12	1	As above	21.77 1.85 9.80	30 6 5#	As above	173 1%	1–60	42	0.18*	(iv)

\# Commencing year 15
* Including surface resistance

2.F.1 Windows — *continued*

No	*Description*		*Capital*	*Cleaning*	*Cyclical*	*Repairs*	*Total*
4	ALUMINIUM	m^2					
a	Standard, white acrylic finish; fixed with lugs or plugged and screwed; average $1.44m^2$						
	(i) fixed light, single glazed		91.28	44.28	11.66	0.29	147.51
	(ii) fixed light; double glazed		108.33	44.28	11.66	0.34	164.61
	(iii) fixed plus fanlight, single glazed		118.95	44.28	22.26	0.37	185.86
	(iv) fixed plus fanlight, double glazed		146.51	44.28	23.02	0.44	214.25
	(v) vertical sliding, single glazed		130.31	44.28	19.38	0.40	194.37
	(vi) vertical sliding, double glazed		213.53	44.28	19.38	0.62	277.81
	(vii) horizontal sliding, single glazed		138.64	44.28	17.38	0.42	200.72
	(viii) horizontal sliding, double glazed		186.30	44.28	17.38	0.55	248.51

Cleaning			Cyclical			Repairs					
Annual Cost	*Frequency No*	*Yr*	*Work*	*Cost*	*Cycle in Years*	*Cause/ Action*	*Cost (£) and Risk*	*Occurs in Years*	*kg/m²*	*R value*	*No*
											4
											a
Clean both sides 2.74	12	1	Clean int. Paint ext	1.85 4.85	6 5#	Replace frame	108 1%	1–60	32	0.17*	(i)
As above 2.74	12	1	As above	1.85 4.85	6 5#	As above	125 1%	1–60	42	0.28*	(ii)
As above 2.74	12	1	Overhaul Clean int. Paint ext.	17.37 1.85 9.80	30 6 5#	Replace frame and/or ironmongery	136 1%	1–60	42	0.16*	(iii)
As above 2.74	12	1	As above	21.77 1.85 9.80	30 6 5#	As above	164 1%	1–60	52	0.23*	(iv)
As above 2.74	12	1	As above	44.37 1.85 4.85	30 6 5#	As above	147 1%	1–60	42	0.16*	(v)
As above 2.74	12	1	As above	44.37 1.85 4.85	30 6 5#	As above	231 1%	1–60	52	0.23*	(vi)
As above 2.74	12	1	As above	32.87 1.85 4.85	30 6 5#	As above	155 1%	1–60	42	0.16*	(vii)
As above 2.74	12	1	As above	32.87 1.85 4.85	30 6 5#	As above	204 1%	1–60	52	0.23*	(viii)

Commencing year 15
* Including surface resistance

No	Description		Capital	Cleaning	Cyclical	Repairs	Total
	2.F.1 Windows — *continued*						
5	UPVC	m²					
a	Standard, BS 2782, reinforced with aluminium alloy; cill; fixed with lugs or plugged and screwed; average 1.44m²						
	(i) fixed light, single glazed		128.55	44.28	7.34	0.39	180.56
	(ii) fixed light, double glazed		140.49	44.28	7.34	0.42	192.53
	(iii) composite fixed/opening, single glazed		192.35	44.28	7.34	0.56	244.53
	(iv) composite fixed/opening, double glazed		208.35	44.28	7.34	0.61	260.58
	(v) tilt and turn, single glazed		160.33	44.28	13.06	0.48	218.15
	(vi) tilt and turn, double glazed		175.60	44.28	13.06	0.72	233.66
6	*ADD* to window, any description, to single glazing or one skin of double glazing	m²					
	(i) toughened clear glass, 5mm		23.88	–	–	–	23.88
	(ii) toughened clear glass, 10mm		73.51	–	–	–	73.51
	(iii) laminated safety glass, clear, 4.4mm		37.48	–	–	–	37.48
	(iv) laminated safety glass, clear, 6.4mm		43.72	–	–	–	43.72
	(v) anti-sun/solar control, 4mm		33.65	–	–	–	33.65
	(vi) anti-sun/solar control, 10mm		138.66	–	–	–	138.66
	2.F.2 External doors See Element 2.H — 'Doors'						

Cleaning			Cyclical			Repairs					
Annual Cost	*Frequency No*	*Yr*	*Work*	*Cost*	*Cycle in Years*	*Cause/ Action*	*Cost (£) and Risk*	*Occurs in Years*	*kg/m²*	*R value*	*No*
											5
											a
Clean both sides 2.74	12	1	Overhaul Clean & polish both sides	6.06 2.75	30 6	Replace frame	145 1%	1–60	11	0.19*	(i)
As above 2.74	12	1	As above	6.06 2.75	30 6	As above	157 1%	1–60	21	0.33*	(ii)
As above 2.74	12	1	As above	6.06 2.75	30 6	Replace frame and/or ironmongery	209 1%	1–60	15	0.20*	(iii)
As above 2.74	12	1	As above	6.06 2.75	30 6	As above	226 1%	1–60	25	0.34*	(iv)
As above 2.74	12	1	As above	38.93 2.75	30 6	As above	177 1%	1–60	25	0.20*	(v)
As above 2.74	12	1	As above	38.93 2.75	30 6	As above	266 1%	1–60	35	0.34*	(vi)
											6
–	–	–	–	–	–	–	–	–	3	–	(i)
–	–	–	–	–	–	–	–	–	15	0.01*	(ii)
–	–	–	–	–	–	–	–	–	1	–	(iii)
–	–	–	–	–	–	–	–	–	6	–	(iv)
–	–	–	–	–	–	–	–	–	–	–	(v)
–	–	–	–	–	–	–	–	–	15	0.10*	(vi)

* Including surface resistance

2.G Internal walls and partitions

No	Description	Capital	Cleaning	Cyclical	Repairs	Total
	NOTE: Add for internal finishings as required					
1	MASONRY m²					
a	Reinforced concrete in situ, 25N; hy reinforcement 13kg/m²; formwork, basic					
	(i) 100 thick	58.19	–	29.28	0.03	87.50
	(ii) 140 thick	65.55	–	34.36	0.03	99.94
	(iii) 240 thick	83.96	–	45.03	0.04	129.03
b	Concrete blocks, dense aggregrate; gauged mortar (1:2:9)					
	(i) 100 thick, solid	18.12	–	10.68	0.01	28.81
	(ii) 140 thick, solid	23.83	–	16.66	0.01	40.50
	(iii) 215 thick, hollow	31.19	–	22.81	0.02	54.02
c	Concrete blocks, lightweight aerated; gauged mortar (1:2:9)					
	(i) 100 thick	16.61	–	11.44	0.01	28.06
	(ii) 140 thick	21.36	–	14.99	0.01	36.36
	(iii) 215 thick	30.51	–	22.50	0.02	53.03
d	Concrete blocks, lightweight aerated, smooth face; gauged mortar (1:2:9); flush pointed					
	(i) 100 thick	21.26	3.23	13.60	0.01	38.10
	(ii) 140 thick	27.28	3.23	19.98	0.01	50.50
	(iii) 215 thick	38.54	3.23	26.21	0.02	68.00

Cleaning			Cyclical			Repairs					
Annual Cost	Frequency No	Yr	Work	Cost	Cycle in Years	Cause/ Action	Cost (£) and Risk	Occurs in Years	kg/m²	R value	No
											1
											a
–	–	–	Form door opening, move partition (one in four)	71.59	20	Defective work/materials	63 0.25%	10–60	210	0.07	(i)
–	–	–	As above	84.02	20	As above	72 0.25%	10–60	294	0.10	(ii)
–	–	–	As above	110.09	20	As above	95 0.25%	10–60	504	0.17	(iii)
											b
–	–	–	As above	26.11	20	As above	22 0.25%	10–60	230	0.06	(i)
–	–	–	As above	40.73	20	As above	29 0.25%	10–60	322	0.09	(ii)
–	–	–	As above	55.76	20	As above	37 0.25%	10–60	295	0.26	(iii)
											c
–	–	–	As above	27.98	20	As above	20 0.25%	10–60	60	0.63	(i)
–	–	–	As above	36.66	20	As above	26 0.25%	10–60	84	0.88	(ii)
–	–	–	As above	55.00	20	As above	37 0.25%	10–60	129	1.34	(iii)
											d
Dust 0.20	6	1	As above	33.24	20	As above	24 0.25%	10–60	60	0.63	(i)
As above 0.20	6	1	As above	48.85	20	As above	32 0.25%	10–60	84	0.88	(ii)
As above 0.20	6	1	As above	64.09	20	As above	46 0.25%	10–60	129	1.34	(iii)

2.G Internal walls and partitions — *continued*

No	Description		Capital	Cleaning	Cyclical	Repairs	Total
1	MASONRY — *continued*	m²					
e	Bricks, common; gauged mortar (1:1:6)						
	(i) half brick wall		23.98	–	16.11	0.01	40.10
	(ii) one brick wall		44.51	–	29.34	0.02	73.87
	(iii) one and a half brick wall		63.21	–	43.21	0.03	106.47
f	Bricks, facings, sand faced (PC £120.00/K), pointed; gauged mortar (1:1:6)						
	(i) extra over common bricks, flemish bond		9.32	3.23	5.36	–	17.91
	(ii) half brick wall, faced both sides, stretcher bond		33.24	6.46	21.45	0.02	61.17
	(iii) half brick wall, faced both sides, flemish bond, snapped headers		41.47	6.46	25.26	0.02	73.21
g	Bricks, facings, machine made (PC £300.00/K), pointed; gauged mortar (1:1:6)						
	(i) extra over common bricks, flemish bond		23.86	3.23	12.09	0.01	39.19
	(ii) half brick wall, faced both sides, stretcher bond		46.16	6.46	27.43	0.02	80.07
	(iii) half brick wall, faced both sides, flemish bond, snapped headers.		49.67	6.46	29.05	0.02	85.20
h	Bricks, facings, hand made (PC £490.00/K), pointed; gauged mortar (1:1:6)						
	(i) extra over common bricks, stretcher bond		38.79	3.23	18.99	0.02	61.03
	(ii) half brick wall, faced both sides, stretcher bond		61.11	6.46	34.34	0.03	101.94
	(iii) half brick wall, faced both sides, flemish bond, snapped headers		65.53	6.46	36.39	0.03	108.41

Cleaning			Cyclical			Repairs					
Annual Cost	*Frequency No*	*Yr*	*Work*	*Cost*	*Cycle in Years*	*Cause/ Action*	*Cost (£) and Risk*	*Occurs in Years*	*kg/m²*	*R value*	*No*
											1
											e
–	–	–	Form door opening, move partition (one in four)	39.40	20	Defective work/ materials	29 0.25%	10–60	192	0.17	(i)
–	–	–	As above	71.73	20	As above	53 0.25%	10–60	392	0.35	(ii)
–	–	–	As above	105.70	20	As above	76 0.25%	10–60	583	0.52	(iii)
											f
Dust 0.20	6	1	As above	13.11	20	As above	9 0.25%	10–60	–	–	(i)
Dust both sides 0.40	6	1	As above	52.45	20	As above	37 0.25%	10–60	192	0.18	(ii)
As above 0.40	6	1	As above	61.5	20	As above	46 0.25%	10–60	192	0.18	(iii)
											g
Dust 0.20	6	1	As above	29.56	20	As above	24 0.25%	10–60	–	–	(i)
Dust both sides 0.40	6	1	As above	67.06	20	As above	51 0.25%	10–60	192	0.18	(ii)
As above 0.40	6	1	As above	71.03	20	As above	54 0.25%	10–60	192	0.18	(iii)
											h
Dust 0.20	6	1	As above	46.44	20	As above	39 0.25%	10–60	–	–	(i)
Dust both sides 0.40	6	1	As above	83.96	20	As above	65 0.25%	10–60	192	0.18	(ii)
As above 0.40	6	1	As above	88.97	20	As above	69 0.25%	10–60	192	0.18	(iii)

2.G Internal walls and partitions – *continued*

No	Description		Capital	Cleaning	Cyclical	Repairs	Total
1	MASONRY – *continued*	m²					
j	Glass blocks, hollow, cross-reeded, clear or similar; cement and sand joints; pointed both sides in mastic						
	(i) 190 × 190 × 80		193.04	6.88	34.54	0.09	234.55
	(ii) 190 × 190 × 80; 6 mm stainless steel rods in joints		298.66	6.88	51.18	0.14	356.86
k	Natural limestone, Cotswold, random rubble; gauged mortar (1:2:9)						
	(i) 275 thick, uncoursed, pointed both sides		61.53	6.46	21.56	0.03	89.58
	(ii) 350 thick; uncoursed, pointed both sides		74.00	6.46	27.45	0.04	107.95
	(iii) 114 thick; coursed, pointed one side		66.13	3.23	9.96	0.03	79.35
	(iv) 150 thick; coursed, pointed one side		74.07	3.23	11.77	0.04	89.11
l	Natural limestone, Portland, ashlar; gauged mortar (1:2:9); bedded against other work; cramps; stone dust pointed one side						
	(i) 50 thick		192.43	3.27	13.67	0.09	209.46
	(ii) 75 thick		263.62	3.27	18.33	0.12	285.34

Cleaning			Cyclical			Repairs					
Annual Cost	*Frequency No*	*Yr*	*Work*	*Cost*	*Cycle in Years*	*Cause/ Action*	*Cost (£) and Risk*	*Occurs in Years*	*kg/m²*	*R value*	*No*
											1
											j
Dust both side 0.40	6	1	Form door opening, move partition (one in four)	84.44	20	Minor defects	199 0.25%	10–60	67	0.12	(i)
Wash both sides 0.18	1	6									
As above 0.40	6	1	As above	125.14	20	As above	305 0.25%	10–60	68	0.12	(ii)
0.18	1	6									
											k
Dust both sides 0.40	6	1	As above	52.72	20	As above	69 0.25%	10–60	600	0.18	(i)
As above 0.40	6	1	As above	67.11	20	As above	84 0.25%	10–60	763	0.23	(ii)
Dust 0.20	6	1	As above	24.34	20	As above	73 0.25%	10–60	250	0.08	(iii)
As above 0.20	6	1	As above	28.78	20	As above	79 0.25%	10–60	327	0.10	(iv)
											l
Dust 0.20	6	1	As above	33.43	20	As above	195 0.25%	10–60	109	0.03	(i)
Specialist clean 0.25	1	30									
As above 0.20	6	1	As above	44.81	20	As above	266 0.25%	10–60	164	0.05	(ii)
0.25	1	30									

2.G Internal walls and partitions — *continued*

No	Description	Capital	Cleaning	Cyclical	Repairs	Total
1	MASONRY — *continued* m²					
m	Reconstructed limestone blocks, Cotswold, ashlar or rough hewn; coloured gauged mortar (1:2:9), pointed one side; 100 thick					
	(i) random, uncoursed	33.84	3.27	21.41	0.02	58.54
	(ii) squared, coursed	36.04	3.27	21.41	0.02	60.74
2	DRY PARTITIONS m²					
a	Softwood, sawn, treated, studding, plates, noggings					
	(i) 38 × 100, 400 centres studs, 600 centres noggings	12.12	–	2.79	0.01	14.92
	(ii) 50 × 100, 400 centres studs, 600 centres noggings	15.98	–	4.37	0.01	20.36
	(iii) 38 × 100, 600 centres studs, 600 centres noggings	9.69	–	2.79	0.01	12.49
	(iv) 50 × 100, 600 centres studs, 600 centres noggings	12.78	–	4.37	0.01	17.16
b	Steel channel studding and plates; sawn softwood fixings; 12.7mm plasterboard lining both sides; joints filled, taped, feathered, for direct decoration					
	(i) 75 thick; one layer plasterboard each side; ½ hour fire resisting	19.91	–	15.85	0.11	35.87
	(ii) 100 thick; two layers plasterboard each side; 1 hour fire resisting	26.57	–	24.58	0.15	51.30
c	Steel channel studding and plates; plain steel lining both sides, enamelled; close fitting joints; glassfibre insulation					
	(i) 50 thick	65.78	3.64	21.88	0.37	91.67
	(ii) 75 thick; ½ hour fire resisting	303.60	3.64	40.97	1.65	349.86

Cleaning			Cyclical			Repairs					
Annual Cost	*Frequency No*	*Yr*	*Work*	*Cost*	*Cycle in Years*	*Cause/ Action*	*Cost (£) and Risk*	*Occurs in Years*	*kg/m²*	*R value*	*No*
											1
											m
Dust 0.20 Specialist clean 0.25	6 1	1 30	Form door opening, move partition (one in four)	52.35	20	Minor defects	37 0.25%	10–60	218	0.07	(i)
As above 0.20 0.25	6 1	1 30	As above	52.35	20	As above	40 0.25%	10–60	218	0.07	(ii)
											2
											a
–	–	–	Form door opening, move partition (one in four)	6.81	20	Dry rot, woodworm	14 0.50%	20–60	11	0.28 (av)	(i)
–	–	–	As above	10.69	20	As above	11 0.50%	20–60	15	0.31 (av)	(ii)
–	–	–	As above	6.81	20	As above	11 0.50%	20–60	9	0.26 (av)	(iii)
–	–	–	As above	10.69	20	As above	17 0.50%	20–60	12	0.29 (av)	(iv)
											b
–	–	–	As above	38.76	20	Replace plasterboard	20 2%	1–60	27	0.34	(i)
–	–	–	As above	60.10	20	As above	28 2%	1–60	52	0.49	(ii)
											c
Dust both sides 0.20 Wash both sides 0.18	6 1	1 6	As above	53.49	20	Replace damaged panels	68 2%	1–60	3	1.32	(i)
As above 0.20 0.18	6 1	1 6	As above	100.16	20	As above	307 2%	1–60	25	2.00	(ii)

2.G Internal walls and partitions — *continued*

No	Description	Capital	Cleaning	Cyclical	Repairs	Total
2	DRY PARTITIONS — *continued* m²					
d	Steel channel studding and plates; stainless steel lining both sides; close fitting joints; glassfibre insulation; 75 thick, ½ hour fire resisting	506.00	3.64	45.29	2.74	557.67
e	Steel channel studding, plates and rails enamelled; 75 thick					
	(i) 6mm, fully glazed	75.90	33.78	17.45	0.43	127.56
	(ii) 6mm, fully double glazed	111.32	33.78	26.81	0.61	172.52
f	Plasterboard prefabricated cellular panel 9.5mm both sides, unfinished; sawn softwood head, cill and edges for fixing; joints scrimmed					
	(i) 57 thick	15.63	–	7.02	0.10	22.75
	(ii) 63 thick	17.24	–	7.86	0.11	25.21
g	Plasterboard panels, laminated in situ, both sides finished for direct decoration, 19 plasterboard core; sawn softwood head and cill					
	(i) 50 thick; 12.7 plasterboard both sides	16.92	–	7.02	0.11	24.05
	(ii) 65 thick; 19 plasterboard both sides	19.75	–	7.86	0.12	27.73
	ADD to the above where appropriate					
h	Glassfibre, flanged, pinned to timber					
	(i) 60 thick	3.20	–	–	0.02	3.22
	(ii) 80 thick	3.77	–	–	0.02	3.79
	(iii) 100 thick	4.29	–	–	0.02	4.31

Cleaning			*Cyclical*			*Repairs*					
Annual Cost	*Frequency No*	*Yr*	*Work*	*Cost*	*Cycle in Years*	*Cause/ Action*	*Cost (£) and Risk*	*Occurs in Years*	*kg/m²*	*R value*	*No*
											2
Dust both side 0.20 Wash both sides 0.18	6 1	1 6	Form door opening, move partition (one in four)	110.73	20	Replace damaged panels	509 2%	1–60	25	2.00	d
											e
Wash both sides 2.09	12	1	As above	42.67	20	Replace damaged glass/rails	79 2%	1–60	18	0.01	(i)
As above 2.09	12	1	As above	65.56	20	As above	114 2%	1–60	33	0.19	(ii)
											f
–	–	–	As above	17.17	20	Replace damaged panels	19 2%	1–60	18	0.30	(i)
–	–	–	As above	19.22	20	As above	20 2%	1–60	24	0.34	(ii)
											g
–	–	–	As above	17.17	20	As above	20 2%	1–60	40	0.28	(i)
–	–	–	As above	19.22	20	As above	23 2%	1–60	52	0.35	(ii)
											h
–	–	–	–	–	–	Defects in structure	3 2%	1–60	2	1.50	(i)
–	–	–	–	–	–	As above	4 2%	1–60	2	2.00	(ii)
–	–	–	–	–	–	As above	4 2%	1–60	3	2.50	(iii)

2.G Internal walls and partitions — *continued*

No	*Description*		*Capital*	*Cleaning*	*Cyclical*	*Repairs*	*Total*
2	DRY PARTITIONS — *continued*	m^2					
	ADD to the above where appropriate — *continued*						
j	Polystyrene, expanded board, fixed with adhesive						
	(i) 25 thick		3.64	–	–	0.03	3.67
	(ii) 50 thick		5.36	–	–	0.04	5.40
k	Terrazzo, precast, white marble aggregate and cement; reinforced						
	(i) 38 thick		116.60	1.54	–	0.67	118.81
	(ii) 50 thick		129.80	1.54	–	0.75	132.09

Cleaning			Cyclical			Repairs					
Annual Cost	*Frequency No*	*Yr*	*Work*	*Cost*	*Cycle in Years*	*Cause/ Action*	*Cost (£) and Risk*	*Occurs in Years*	*kg/m²*	*R value*	*No*
											2
											j
–	–	–	–	–	–	Defects in structure	6 2%	1–60	1	0.86	(i)
–	–	–	–	–	–	As above	7 2%	1–60	2	1.35	(ii)
											k
Dust both sides 0.07 Wash both sides 0.16	6 1	1 6	–	–	–	Damaged fixings	125 2%	1–60	93	0.02	(i)
As above 0.07 0.18	6 1	1 6	–	–	–	As above	139 2%	1–60	122	0.03	(ii)

2 SUPERSTRUCTURE

2.H Doors

No	Description	Capital	Cleaning	Cyclical	Repairs	Total
1	SOFTWOOD (762 × 1981) Nr					
a	Matchboard 19, ledges and braces 26; frame 94 × 57; quadrant; butts; lock and levers; painted	181.64	95.35	78.46	0.55	356.00
b	Framed 44, ledges and braces 26, matchboarding 19; frame 94 × 57; quadrant; butts; lock and levers; painted	195.17	95.35	78.46	0.56	369.54
2	STANDARD FLUSH (762 × 1981) Nr					
a	Skeleton core, 35 or 40; lining 144 × 32; stop 44 × 13; architraves; butts; lock and levers					
	(i) hardboard faced; painted	162.09	95.35	63.77	0.50	321.71
	(ii) plywood faced; painted	168.86	95.35	63.77	0.51	328.49
	(iii) sapele veneered; varnished	173.98	129.45	63.54	0.53	367.50
b	Skeleton core, 44; frame 94 × 57; quadrant; weatherboard; butts; lock and levers; plywood faced; painted	197.55	95.35	80.13	0.59	373.62
c	Solid core, 44, ½ hour fire (30/20); lining 144 × 32; stop 44 × 25; architraves; butts, lock and levers; overhead closer					
	(i) hardboard faced; painted	267.19	95.35	141.72	0.80	505.06

Cleaning			Cyclical			Repairs					
Annual Cost	*Frequency No*	*Yr*	*Work*	*Cost*	*Cycle in Years*	*Cause/ Action*	*Cost (£) and Risk*	*Occurs in Years*	*kg*	*R value*	*No*
											1
Dust 5.61 Wash 0.29	50 2	1 1	Repaint Ironmongery Ease door	15.91 14.94 23.27	5 10 15	Refix frame, repair door/ ironmongery	204 1%	1–60	52	0.33 §*	a
As above 5.61 0.29	50 2	1 1	As above	15.91 14.94 23.27	5 10 15	As above	209 1%	1–60	66	0.33 §*	b
											2
											a
Dust 5.61 Wash 0.29	50 2	1 1	Repaint Ironmongery Ease door	15.91 14.94 23.27	6 10 20	Refix frame, repair door/ ironmongery	184 1%	1–60	62	0.52*	(i)
As above 5.61 0.29	50 2	1 1	As above	15.91 14.94 23.27	6 10 20	As above	191 1%	1–60	58	0.51*	(ii)
Dust and clean 8.01	50	1	Revarnish Ironmongery Ease door	14.59 14.59 31.11	6 10 20	As above	196 1%	1–60	59	0.51*	(iii)
Dust 5.61 Wash 0.29	50 2	1 1	Repaint Ironmongery Ease door	16.15 14.94 23.27	5 10 15	As above	220 1%	1–60	58	0.48 §*	b
											c
As above 5.61 0.29	50 2	1 1	As above	15.91 80.11 23.27	6 10 20	As above	297 1%	1–60	91	0.58*	(i)

§ One side exterior
* Including surface resistance

2.H Doors—*continued*

No	Description	Capital	Cleaning	Cyclical	Repairs	Total
2	STANDARD FLUSH (762 × 1981) — *continued* Nr					
c	Solid core, 44, ½ hour fire (30/20); lining 144 × 32; stop 44 × 25; architraves; butts, lock and levers; overhead closer — *continued*					
	(ii) plywood faced; painted	274.93	95.35	141.72	0.82	512.82
	(iii) sapele veneered; varnished	295.26	129.45	149.11	0.86	574.68
d	Solid core, 44, ½ hour fire (30/30); lining 144 × 32; stop 44 × 25; architraves; butts; lock and levers; overhead closer					
	(i) sapele veneered; varnished	297.28	129.45	149.11	0.86	576.70
	(ii) melamine laminate veneered	405.39	26.67	127.87	1.16	561.09
e	Solid core, 54, one hour fire (60/45); frame 144 × 57; architraves; butts, locks and levers; overhead closer; interior					
	(i) plywood faced; painted	363.84	95.35	141.72	1.07	601.98
	(ii) sapele veneered; varnished	408.06	129.45	149.11	1.19	687.81

Cleaning			Cyclical			Repairs					
Annual Cost	Frequency No	Yr	Work	Cost	Cycle in Years	Cause/ Action	Cost (£) and Risk	Occurs in Years	kg	R value	No
											2
											c
Dust 5.61 Wash 0.29	50 2	1 1	Repaint Ironmongery Ease door	15.91 80.11 23.27	6 10 20	Refix frame, repair door/ ironmongery	303 1%	1–60	87	0.57*	(ii)
Dust and clean 8.01	50	1	Revarnish Ironmongery Ease door	14.59 86.14 31.11	6 10 20	As above	319 1%	1–60	88	0.57*	(iii)
											d
As above 8.01	50	1	As above	14.59 86.14 31.11	6 10 20	As above	319 1%	1–60	88	0.57*	(i)
Wash 1.65	50	1	Repaint lining Ironmongery Ease door	5.30 86.14 31.11	6 10 20	As above	429 1%	1–60	105	0.57*	(ii)
											e
Dust 5.61 Wash 0.29	50 2	1 1	Repaint Ironmongery Ease door	15.91 80.11 23.27	6 10 20	As above	396 1%	1–60	97	0.62*	(i)
Dust and clean 8.01	50	1	Revarnish Ironmongery Ease door	14.59 86.14 31.11	6 10 20	As above	440 1%	1–60	98	0.62*	(ii)

* Including surface resistance

2.H Doors—*continued*

No	Description	Capital	Cleaning	Cyclical	Repairs	Total
2	STANDARD FLUSH (762 × 1981) – *continued* Nr					
f	Solid core, 54, one hour fire (60/60); frame 144 × 57; architraves; butts; lock and levers; overhead closer; intumescent strip					
	(i) sapele veneered; varnished	427.79	129.45	149.11	1.24	707.59
	(ii) melamine laminate veneered	497.83	26.67	127.87	1.45	653.82
3	*ADD* to door, any description for Nr					
a	opening in door; 6mm wired glass; beads, screws and cups; 0.5m^2	37.17	21.33	–	0.11	58.61
b	where doors in pairs; add push plates; kicking plates; pull handles; bolts; deduct lining, etc; per pair	48.64	–	–	0.15	48.79
c	as last and with floor springs to each leaf; deduct overhead closers; per pair	421.05	–	283.16	1.24	705.45
d	fanlight, fixed; 6mm wired glass; beads, screws and cups; 0.5m^2	62.34	21.33	–	0.19	83.86
4	SOFTWOOD – NON STANDARD Nr					
a	Panelled, 44; lining 144 × 32; stop 44 × 26; architraves; butts; lock and levers; painted; (2m^2)					
	(i) four panels, 19 ply, mouldings both sides	244.26	95.35	63.77	0.74	404.12
	(ii) six panels, raised and fielded mouldings both sides	280.47	95.35	63.77	0.83	440.42

Cleaning			*Cyclical*			*Repairs*					
Annual Cost	*Frequency No*	*Yr*	*Work*	*Cost*	*Cycle in Years*	*Cause/ Action*	*Cost (£) and Risk*	*Occurs in Years*	*kg*	*R value*	*No*
											2
											f
Dust and clean 8.01	50	1	Revarnish Ironmongery Ease door	14.59 86.14 31.11	6 10 20	Refix frame, repair door/ ironmongery	462 1%	1–60	98	0.62*	(i)
Wash 1.65	50	1	Repaint lining Ironmongery Ease door	5.30 86.14 31.11	6 10 20	As above	539 1%	1–60	115	0.62*	(ii)
											3
Clean glass 1.32	12	1	–	–	–	Refix beads	40 1%	1–60	–	–	a
–	–	–	–	–	–	Additional easing	55 1%	1–60	–	–	b
–	–	–	Replace floor springs	236.76	10	As above	462 1%	1–60	–	–	c
Clean glass 1.32	12	1	–	–	–	Refix beads	69 1%	1–60	19	0.25*	d
											4
											a
Dust 5.61 Wash 0.29	50 2	1 1	Repaint Ironmongery Ease door	15.91 14.94 23.27	6 10 20	Refix frame, repair door/ ironmongery	275 1%	1–60	70	0.49*	(i)
As above 5.61 0.29	50 2	1 1	As above	15.91 14.94 23.27	6 10 20	As above	308 1%	1–60	68	0.49*	(ii)

* Including surface resistance

No	2.H Doors—*continued* Description	Capital	Cleaning	Cyclical	Repairs	Total
4	SOFTWOOD – NON STANDARD – *continued* Nr					
a	Panelled, 44; lining 144 × 32; stop 44 × 26; architraves; butts; lock and levers; painted; (2m²) – *continued*					
	(iii) two panels; one panel 6mm wired glass	242.71	127.35	63.77	0.74	434.57
	(iv) two panels; two panels 6mm wired glass	259.59	127.35	63.77	0.77	451.48
5	HARDWOOD – NON STANDARD Nr					
a	Panelled 44; lining 144 × 32; stop 44 × 26; architraves; butts; lock and levers; varnished (2m²)					
	(i) four panels, 19 ply, mouldings both sides					
	(1) African Mahogany	414.92	129.45	63.96	1.19	609.52
	(2) Afrormosia	503.23	129.45	63.96	1.42	698.06
	(ii) six panels, raised and fielded, mouldings both sides					
	(1) African Mahogany	473.98	129.45	63.96	1.36	668.75
	(2) Afrormosia	576.28	129.45	63.96	1.66	771.35
	(iii) two panels; one panel 6mm wired glass					
	(1) African Mahogany	376.68	161.45	63.96	1.10	603.19

Cleaning			Cyclical			Repairs					
Annual Cost	*Frequency No*	*Yr*	*Work*	*Cost*	*Cycle in Years*	*Cause/ Action*	*Cost (£) and Risk*	*Occurs in Years*	*kg*	*R value*	*No*
											4
											a
Dust 5.61	50	1	Repaint	15.91	6	Refix frame, repair door/ ironmongery	275 1%	1–60	64	0.37*	(iii)
Wash 0.29	2	1	Ironmongery	14.94	10						
Clean glass 1.98	12	1	Ease door	23.27	20						
As above 5.61	50	1	As above	15.91	6	As above	286 1%	1–60	60	0.39*	(iv)
0.29	2	1		14.94	10						
1.98	12	1		23.27	20						
											5
											a
											(i)
Dust and clean 8.01	50	1	Revarnish	14.59	6	Refix frame, repair door/ ironmongery	440 1%	1–60	93	0.43*	(1)
			Ironmongery	14.94	10						
			Ease door	31.11	20						
As above 8.01	50	1	As above	14.59	6	As above	528 1%	1–60	97	0.43*	(2)
				14.94	10						
				31.11	20						
											(ii)
As above 8.01	50	1	As above	14.59	6	As above	506 1%	1–60	92	0.43*	(1)
				14.94	10						
				31.11	20						
As above 8.01	50	1	As above	14.59	6	As above	616 1%	1–60	96	0.43*	(2)
				14.94	10						
				31.11	20						
											(iii)
Dust and clean 8.01	50	1	As above	14.59	6	As above	407 1%	1–60	82	0.35*	(1)
				14.94	10						
Clean glass 1.98	12	1		31.11	20						

* Including surface resistance

2.H Doors — *continued*

No	Description		Capital	Cleaning	Cyclical	Repairs	Total
5	HARDWOOD – NON STANDARD — *continued*	Nr					
a	Panelled, 44; lining 144 × 32; stop 44 × 26; architraves; butts; lock and levers; varnished — *continued*						
	(iii) two panels; one panel 6mm wired glass — *continued* (2) Afrormosia		463.53	161.45	63.96	1.33	690.27
	(iv) two panels; two panels 6mm wired glass (1) African Mahogany		404.83	193.45	63.96	1.16	663.40
	(2) Afrormosia		510.74	193.45	63.96	1.45	769.60
6	ROLLER SHUTTERS	Nr					
a	Hardwood laths, steel roller; hardwood frame; plywood hood; painted; 2150 × 2150; pole and hook operation; (4.6m²)		884.47	190.70	854.50	1.26	1930.93
b	Galvanised steel laths, roller, guides; hood; painted						
	(i) 5500 × 4250; endless hand chain operation; (23.4m²)		3103.21	889.18	2958.09	4.30	6954.78
	(ii) 4600 × 3700; crank handle operation; (17.0 m²)		2484.98	317.40	2329.41	3.48	5135.27
	(iii) 6100 × 6100; electric motor operation; (37.2m²)		6023.41	1333.93	5269.44	8.22	12635.00
c	Aluminium laths, guides, hood; steel roller; 3600 × 3050; endless hand chain operation; (11.0m²)		1905.87	381.08	1800.51	2.67	4090.13

Cleaning			Cyclical			Repairs					
Annual Cost	Frequency No	Yr	Work	Cost	Cycle in Years	Cause/ Action	Cost (£) and Risk	Occurs in Years	kg	R value	No
											5
											a
Dust and clean 8.01	50	1	Revarnish	14.59	6	Refix frame, repair door/ ironmongery	495 1%	1–60	86	0.35*	(iii) (2)
			Ironmongery	14.94	10						
Clean glass 1.98	12	1	Ease door	31.11	20						
As above 8.01	50	1	As above	14.59	6	As above	429 1%	1–60	74	0.36*	(iv) (1)
				14.94	10						
3.96	12	1		31.11	20						
As above 8.01	50	1	As above	14.59	6	As above	539 1%	1–60	77	0.36*	(2)
				14.94	10						
3.96	12	1		31.11	20						
											6
Dust 11.22	50	1	Replace	896.60	20	Repair links, repair gear, refix frame	935 0.50%	1–60	60	0.23 §*	a
Wash			Repaint	33.00	5						
0.58	2	1	Maintain gear	24.39	1						
As above 52.36	50	1	As above	3163.88	30	As above	3190 0.50%	1–60	353	0.18 §*	b (i)
2.66	2	1		154.00	5						
				121.94	1						
As above 18.70	50	1	As above	2533.51	30	As above	2585 0.50%	1–60	273	0.18 §*	(ii)
0.94	2	1		110.00	5						
				97.55	1						
As above 78.54	50	1	As above	6132.61	30	As above	6105 0.50%	1–60	567	0.18 §*	(iii)
4.00	2	1		231.00	5						
				219.48	1						
Above 22.44	50	1	Replace	1941.61	25	As above	1980 0.50%	1–60	89	0.18 §*	c
1.14	2	1	Degrease	26.64	6						
			Maintain gear	73.16	1						

§ One side exterior
* Including surface resistance

2.H Doors — *continued*

No	*Description*	*Capital*	*Cleaning*	*Cyclical*	*Repairs*	*Total*
7	METAL DOORS Nr					
a	Galvanised steel door and frame; fully glazed; butts; lock and levers					
	(i) 1000 × 2060; painted; ($2m^2$)	514.07	175.99	73.33	1.48	764.87
	(ii) 1000 × 2060; polyester powder coated; ($2m^2$)	602.55	175.99	39.61	1.72	819.87
b	Aluminium door and frame; acrylic coated					
	(i) 860 × 2060; fully glazed; butts; lock and levers; ($1.8m^2$)	404.18	154.66	38.15	1.19	598.18
	(ii) 860 × 2060; fully double glazed; butts; lock and levers; ($1.8m^2$)	569.98	290.74	38.15	1.63	900.50
	(iii) 2450 × 2100; sliding; fully double glazed; lock; hardwood sub-frame; ($5.1m^2$)	838.68	440.06	83.86	2.37	1364.97
c	UPVC covered aluminium or steel cored door and frame; 1000 × 2060; fully double glazed; butts; lock and levers; fully double glazed; ($2m^2$)	691.63	175.99	44.69	1.96	914.27
8	GARAGE DOORS (2134 × 2134) Nr					
a	Standard pair, ledged, framed, braced, matchboarded; side hung; matching frame; butts; lock and handles; bolts; decorated					
	(i) softwood	334.65	190.70	163.34	1.01	689.70
	(ii) western red cedar	522.80	190.70	163.34	1.51	878.35
	(iii) oak	1642.12	190.70	174.67	4.53	2012.02

Cleaning			Cyclical			Repairs					
Annual Cost	*Frequency No*	*Yr*	*Work*	*Cost*	*Cycle in Years*	*Cause/ Action*	*Cost (£) and Risk*	*Occurs in Years*	*kg*	*R value*	*No*
											7
											a
Dust 5.61 Clean glass 5.28	50 12	1 1	Repaint Overhaul	16.50 22.17	5 10	Refix frame, repair door/ ironmongery	550 1%	1–60	42	0.18§*	(i)
As above 5.61 5.28	50 12	1 1	Repaint external Overhaul	8.55 22.17	5# 10	As above	638 1%	1–60	42	0.18§*	(ii)
											b
As above 5.61 3.96	50 12	1 1	As above	5.70 24.60	5# 10	As above	440 1%	1–60	33	0.17§*	(i)
As above 14.03 3.96	50 12	1 1	As above	5.70 24.60	5# 10	As above	605 1%	1–60	53	0.28§*	(ii)
As above 14.03 13.20	50 12	1 1	As above	16.34 49.20	5# 10	As above	880 1%	1–60	162	0.28§*	(iii)
As above 5.61 5.28	50 12	1 1	Repair surface treatment external Overhaul	15.20 22.17	10 10	As above	726 1%	1–60	62	0.28§*	c
											8
											a
Dust 11.22 Wash 0.58	50 2	1 1	Redec. Overhaul	37.96 46.53	5 10	Refix frame, repair door/ ironmongery	374 1%	1–60	81	0.33§*	(i)
As above 11.22 0.58	50 2	1 1	As above	37.96 46.53	5 10	As above	561 1%	1–60	64	0.33§*	(ii)
As above 11.22 0.58	50 2	1 1	As above	35.34 62.22	5 10	As above	1683 1%	1–60	115	0.30§*	(iii)

* Including surface resistance
§ One side exterior
\# Commencing year 15

2.H Doors – *continued*

No	Description		Capital	Cleaning	Cyclical	Repairs	Total
8	GARAGE DOORS (2134 × 2134) – *continued*	Nr					
b	Standard one piece, ledged, framed, braced, matchboarded; up and over; matching frame; swinging gear; lock and handles; decorated						
	(i) softwood		459.86	190.70	125.78	1.33	777.67
	(ii) western red cedar		650.01	190.70	125.78	1.84	968.33
	(iii) oak		1781.95	190.70	123.86	4.92	2101.43
c	Standard one piece, pressed galvanised steel moulded door; up and over; softwood frame; swinging gear; lock and handles; decorated		347.66	190.70	145.90	1.01	685.27
9	FLEXIBLE DOORS	Nr					
a	Standard industrial doors in two leaves; pivotting units with springs; top and bottom plates; steel tube frame; painted						
	(i) 6mm clear rubber doors; 43 frame; 2440 × 2440; ($6m^2$)		680.42	204.76	1270.12	1.90	2157.20
	(ii) 8mm rubber doors with perspex panels; 43 frame; 2440 × 2440; ($6m^2$)		880.85	204.76	1177.39	2.49	2265.49
	(iii) 9mm clear rubber doors; 63 frame; 2740 × 3350; ($9.2m^2$)		1166.74	307.22	949.01	3.26	2426.23
	(iv) 12mm rubber doors with clear panels; 63 frame; 2740 × 3350; ($9.2m^2$)		1588.39	307.22	1048.54	4.41	2948.56
b	Standard industrial doors in two leaves; pivotting units with springs; aluminium frame; clear PVC doors; 2400 × 2400; ($5.8m^2$)		804.54	204.76	1426.16	2.25	2437.71

Cleaning			Cyclical			Repairs					
Annual Cost	Frequency No	Yr	Work	Cost	Cycle in Years	Cause/ Action	Cost (£) and Risk	Occurs in Years	kg	R value	No
											8
											b
Dust 11.22 Wash 0.58	50 2	1 1	Redec. Maintain gear Overhaul	27.24 6.07 14.94	5 3 10	Refix frame, repair door/ ironmongery	495 1%	1–60	63	0.33§*	(i)
As above 11.22 0.58	50 2	1 1	As above	27.24 6.07 14.94	5 3 10	As above	682 1%	1–60	49	0.33§*	(ii)
As above 11.22 0.58	50 2	1 1	As above	25.43 6.07 17.63	5 3 10	As above	1826 1%	1–60	93	0.30§*	(iii)
As above 11.22 0.58	50 2	1 1	As above	33.20 61.07 17.63	5 3 10	As above	378 1%	1–60	60	0.18§*	c
											9
											a
Wash 12.67	12	1	Replace leaves Repaint frame	440.00 7.70	5 5	Refix leaf, replace spring/pivot	704 1%	1–60	55	0.28§*	(i)
As above 12.67	12	1	As above	605.00 7.70	7 5	As above	924 1%	1–60	65	0.29§*	(ii)
As above 19.01	12	1	As above	770.00 9.90	10 5	As above	1210 1%	1–60	108	0.30§*	(iii)
As above 19.01	12	1	As above	1100.00 9.90	12 5	As above	1639 1%	1–60	133	0.31§*	(iv)
Wash 12.67	12	1	As above	495.00 7.70	5 5	As above	836 1%	1–60	93	0.30§*	b

* Including surface resistance
§ One side exterior

2.H Doors – *continued*

No	Description		Capital	Cleaning	Cyclical	Repairs	Total
9	FLEXIBLE DOORS – *continued*	Nr					
c	Strip curtain; 2mm clear PVC; 80 × 80 × 6 painted steel overhead fixing; 1800 × 2100; (3.8m^2)		133.16	170.66	650.98	0.41	955.21

Cleaning			Cyclical			Repairs					
Annual Cost	*Frequency No*	*Yr*	*Work*	*Cost*	*Cycle in Years*	*Cause/ Action*	*Cost (£) and Risk*	*Occurs in Years*	*kg*	*R value*	*No*
											9
Wash 10.56	12	1	Replace strips Repaint frame	82.50 2.20	2 5	Refix strips, replace fixings	154 1%	1–60	32	0.25§* theoretically	c

* Including surface resistance
§ One side exterior

3 INTERNAL FINISHES

3.A Wall finishes

No	Description	Capital	Cleaning	Cyclical	Repairs	Total
1	WALLS m^2					
a	Extra over brickwork for fair face and pointing as the work proceeds	2.86	5.49	–	0.01	8.36
b	Extra over common brickwork for facing brickwork (PC £250.00/K) pointed, stretcher bond	15.98	5.49	–	0.01	21.48
c	Lightweight plaster 13, two coat	5.89	See decoration	–	0.04	5.93
d	Cement and sand or dense plaster base 10, gypsum finish 5	6.68	As above	–	0.05	6.73
e	Emulsion paint matt/vinyl, mist and two coats	1.89	5.79	3.22	0.05	10.95
f	Paint oil/eggshell, prime and two coats	3.28	5.79	5.28	0.05	14.40
g	Textured plastic finish, patterned	2.83	5.79	1.93	0.08	10.63
h	Vinyl paper (PC £4.20/roll)	4.02	5.49	6.97	0.16	16.64
j	Tiles, ceramic, cushion edged, standard colour, 108 × 108 × 4; backing 15	29.16	110.70	–	0.43	140.29
k	Tiles eg. slate, cleft or honed (PC £37.00/m^2); backing 12	52.13	18.42	–	0.43	70.98
l	Tiles, glass mosaic; backing 13	67.71	0.23	–	1.00	68.94

Cleaning			Cyclical			Repairs					
Annual Cost	*Frequency No*	*Yr*	*Work*	*Cost*	*Cycle in Years*	*Cause/ Action*	*Cost (£) and Risk*	*Occurs in Years*	*kg/m²*	*R value*	*No*
											1
Dust 0.34	6	1	–	–	–	Repair pointing	3 2%	10–60	–	–	a
As above 0.34	6	1	–	–	–	As above	3 2%	10–60	–	–	b
See decoration	–	–	–	–	–	Repair cracks	8 2%	1–60	6	0.08	c
As above	–	–	–	–	–	As above	10 2%	1–60	21	0.05	d
Dust 0.34 Wash 0.13	6 1	1 6	Repaint	1.41	6	–	2 10%	1–60	–	–	e
As above 0.34 0.13	6 1	1 6	As above	2.31	6	–	2 10%	1–60	–	–	f
As above 0.34 0.13	6 1	1 6	Resurface	3.74	18	–	3 10%	1–60	3	0.01	g
Dust 0.34	6	1	Replace	5.83	10	Patch, refix	6 10%	1–60	1	0.01	h
Wash 6.85	12	1	–	–	–	Loosened/ damaged tiles	32 5%	1–60	31	0.03	j
As above 1.14	2	1	–	–	–	As above	32 5%	1–60	43	0.03	k
As above 0.10	1	6	–	–	–	As above	74 5%	1–60	31	0.03	l

3.A Wall finishes—*continued*

No	Description	Capital	Cleaning	Cyclical	Repairs	Total
1	WALLS — *continued* m²					
m	Dry lining, plasterboard 12, joints taped, feathered; 25 × 50 sawn sofwood bearers, 600 centres, to masonry	9.86	See decoration	–	0.07	9.93
n	Dry lining, plasterboard 12, plaster skim coat; 25 × 50 sawn softwood bearers, 600 centres, to masonry	12.40	As above	–	0.08	12.48
p	Dry lining, plasterboard 12, joints taped, feathered; 50 metal firrings, 600 centres, to masonry	14.21	As above	–	0.09	14.30
q	Dry lining, plasterboard 12, plaster skim coat; 50 metal firrings, 600 centres, to masonry	16.74	As above	–	0.10	16.84
r	Dry lining; melamine faced chipboard 12, open joints; metal system	25.17	18.42	–	0.14	43.73
s	Dry lining, plasterboard 12, plastic faced; PVC trim to joints; 25 × 50 sawn softwood bearers, 600 centres, to masonry	14.23	18.42	–	0.09	32.74
t	Dry lining, thermal board 25; fixed with mortar dabs	7.95	See decoration	–	0.05	8.00
u	Dry lining, wrought softwood 19, matchboarding, t&g and V joints; clear finish, 2 coats polyurethane; 25 × 50 sawn softwood bearers, 600 centres, to masonry	22.54	5.79	4.71	0.07	33.11
v	Dry lining, wrought hardwood, 19, sapele, boarding, t&g and V joints, diagonal/chevron; clear finish, 2 coats polyurethane; 25 × 50 sawn softwood bearers, 600 centres, to masonry	45.31	5.79	4.71	0.13	55.94
w	Dry lining, plywood 6, hardwood decorative veneer; 25 × 50 sawn softwood bearers, 600 centres, to masonry	14.84	5.79	–	0.05	20.68
2	*ADD* to Items 1m to 1w where appropriate					
a	Insulation 25, between battens at 600 centres, fixed with adhesive	4.68	–	–	–	4.68
b	Foil back to plasterboard	0.47	–	–	–	0.47
c	Vapour barrier behind dry lining to external wall	1.84	–	–	–	1.84

Cleaning			Cyclical			Repairs					
Annual Cost	Frequency No	Yr	Work	Cost	Cycle in Years	Cause/ Action	Cost (£) and Risk	Occurs in Years	kg/m²	R value	No
											1
See dec-oration	–	–	–	–	–	Repair cracks	13 2%	1–60	13	0.26	m
As above	–	–	–	–	–	As above	14 2%	1–60	16	0.26	n
As above	–	–	–	–	–	As above	17 2%	1–60	17	0.26	p
As above	–	–	–	–	–	As above	19 2%	1–60	20	0.26	q
Wash 1.14	2	1	–	–	–	Repair	26 2%	1–60	16	0.27	r
As above 1.14	2	1	–	–	–	Refix trim	17 2%	1–60	13	0.26	s
See dec-oration	–	–	–	–	–	Repair joints	10 2%	1–60	12	0.43	t
Dust 0.34 Wash 0.13	6 1	1 6	Redec	2.06	6	Dry rot, wood worm	25 1%	1–60	11	0.32	u
As above 0.34 0.13	6 1	1 6	As above	2.06	6	As above	50 1%	1–60	15	0.30	v
As above 0.34 0.13	6 1	1 6	–	–	–	As above	19 1%	1–60	4	0.22	w
											2
–	–		–	–	–	–	–	–	1	0.41	a
–	–		–	–	–	–	–	–	1	0.16	b
–	–		–	–	–	–	–	–	1	0.01	c

3.B Floor finishes

No	Description	Capital	Cleaning	Cyclical	Repairs	Total
1	FLOORS m²					
a	Paint; primer and two coats	3.34	50.10	9.47	0.11	63.02
b	Cement and sand (1:3), trowelled, 50, incorporated hardener	9.43	50.10	–	0.18	59.71
c	Granolithic finish, non-slip; cement and sand base, monolithic; 50 total	16.17	50.10	–	0.27	66.54
d	Terrazzo finish, cement and sand base, monolithic; 50 total	41.82	50.10	–	0.36	92.28
e	Resinous finish; cement and sand base; 50 total	22.15	50.10	–	0.38	72.63
f	Urethane 6, direct to concrete (Prodorcrete GT)	40.43	50.10	–	0.24	90.77
g	Asphalt BS 1076, black, 25; felt underlay; cement and sand screed 25	18.83	50.10	–	0.46	69.39
h	Quarry tiles 12; cement and sand screed; 50 total	26.05	50.10	–	0.40	76.55
j	Ceramic tiles 10; cement and sand screed; 50 total	31.10	50.10	–	0.46	81.66
k	Granite tiles, black, polished, 20; cement and sand screed; 50 total	253.00	50.10	–	0.73	303.83
l	Terrazzo tiles, coloured, 20; cement and sand screed; 50 total	47.82	50.10	–	0.41	98.33

Cleaning			Cyclical			Repairs					
Annual Cost	*Frequency No*	*Yr*	*Work*	*Cost*	*Cycle in Years*	*Cause/ Action*	*Cost (£) and Risk*	*Occurs in Years*	*kg/m²*	*R value*	*No*
											1
Sweep 2.38 Scrub/ wash 0.72	250 50	1 1	Re-paint two coats	4.14	6	Flaking off damp surface	4 10%	1–60	–	–	a
As above 2.38 0.72	250 50	1 1	–	–	–	Cracking, surface wear	13 5%	1–60	79	0.09	b
As above 2.38 0.72	250 50	1 1	–	–	–	As above	20 5%	1–60	79	0.09	c
As above 2.38 0.72	250 50	1 1	–	–	–	Cracking	45 3%	1–60	85	0.09	d
As above 2.38 0.72	250 50	1 1	–	–	–	As above	28 5%	1–60	83	0.09	e
As above 2.38 0.72	250 50	1 1	–	–	–	Cracking, surface wear	44 2%	1–60	90	0.01	f
As above 2.38 0.72	250 50	1 1	–	–	–	Surface wear	17 10%	1–60	100	0.08	g
As above 2.38 0.72	250 50	1 1	–	–	–	Loosened/ damaged tiles	30 5%	1–60	83	0.09	h
As above 2.38 0.72	250 50	1 1	–	–	–	As above	34 5%	1–60	82	0.09	j
As above 2.38 0.72	250 50	1 1	–	–	–	As above	270 1%	1–60	100	0.06	k
As above 2.38 0.72	250 50	1 1	–	–	–	As above	51 3%	1–60	93	0.07	l

3.B Floor finishes—*continued*

No	*Description*		*Capital*	*Cleaning*	*Cyclical*	*Repairs*	*Total*
1	FLOORS — *continued*	m^2					
m	Marble 20, Italian; cement and sand screed; 50 total		184.47	50.10	–	1.01	235.58
n	Hardwood blocks, Iroko, 275 × 75 × 25, t&g, bedded in mastic, to pattern; cement and sand screed; 50 total; sanded, sealed, wax polished		51.29	50.10	27.15	0.14	128.68
p	Hardwood block slips, felt backing 5, up to 100 × 50, bedded in adhesive; cement and sand screed; 50 total		32.41	50.10	31.60	0.10	114.21
q	Linoleum sheet 3, fixed with adhesive; cement and sand screed; 50 total		20.56	50.10	9.87	0.31	80.84
r	Vinyl tiles 2, fixed with adhesive; cement and sand screed; 50 total		18.06	50.10	8.21	0.28	76.65
s	Vinyl sheet 2.5, fixed with adhesive, seamed joints; cement and sand screed; 50 total		19.93	50.10	9.45	0.31	79.79
t	Thermoplastic tiles 2, fixed with adhesive; cement and sand screed; 50 total		13.92	50.10	5.46	0.24	69.72
u	Rubber safety floor, studded, 4, fixed with adhesive; cement and sand screed; 50 total		34.30	61.57	11.70	0.50	108.07
v	Cork tiles 4, fixed with adhesive; cement and sand screed; 50 total; sealed		18.80	230.46	16.01	0.30	265.57

Cleaning			Cyclical			Repairs					
Annual Cost	Frequency No	Yr	Work	Cost	Cycle in Years	Cause/ Action	Cost (£) and Risk	Occurs in Years	kg/m²	R value	No
											1
Sweep 2.38 Scrub/ wash 0.72	250 50	1 1	–	–	–	Loosened/ damaged tiles	187 2%	1–60	47	0.07	m
As above 2.38 0.72	250 50	1 1	Wax polish Sand, seal, wax polish	1.65 2.77	1 30	Loosened blocks, surface wear	51 1%	1–60	57	0.21	n
As above 2.38 0.72	250 50	1 1	Wax polish Replace entirely	1.65 28.37	1 30	Loosened slips, surface wear	36 1%	1–60	70	0.11	p
As above 2.38 0.72	250 50	1 1	Replace lino, patch screed	14.86	15	Patch lino, patch screed	23 5%	1–60	77	0.10	q
As above 2.38 0.72	250 50	1 1	Replace tiles, patch screed	12.36	15	Patch tiles, patch screed	21 5%	1–60	76	0.09	r
As above 2.38 0.72	250 50	1 1	Replace sheet, patch screed	14.23	15	Patch sheet, patch screed	23 5%	1–60	78	0.10	s
As above 2.38 0.72	250 50	1 1	Replace tiles, patch screed	8.22	15	Patch tiles, patch screed	18 5%	1–60	77	0.10	t
Wash 3.81	250	1	Replace sheet, patch screed	28.60	20	Patch sheet, patch screed	37 5%	1–60	79	0.10	u
Sweep 2.38 Scrub/ wash 11.88	250 12	1 1	Replace tiles, patch screed	13.39	10	Patch tiles, patch screed	27 5%	1–60	75	0.14	v

3.B Floor finishes—*continued*

No	*Description*	*Capital*	*Cleaning*	*Cyclical*	*Repairs*	*Total*
1	FLOORS — *continued* m²					
w	Foam backed vinyl; fixed with adhesive; cement and sand screed; 50 total	19.13	50.10	8.79	0.12	78.14
x	Carpet tiles, needle punch, fixed with adhesive, cement and sand screed; 50 total	16.50	15.35	13.67	0.32	45.84
y	Carpet, needlepunch, fixed with adhesive, cement and sand screed; 50 total	16.09	15.35	12.90	0.31	44.65
z	Carpet, broadloom (PC £14.00/m²); grippers at edges; underlay; cement and sand screed; 50 total	39.78	15.61	38.19	0.63	94.21
aa	Raised floor system; 19 chipboard, t&g; sealed; 50; wrought softwood bearers with insulation strips under, 400 centres	18.06	See covering	–	0.11	18.17
ab	Chipboard, 19, t&g; sealed	6.89	As above	–	0.05	6.94
ac	Wrought softwood boards, 25, t&g	16.31	As above	–	0.11	16.42
ad	Wrought hardwood boards, 25, Iroko, t&g; sealed and wax polished	39.68	50.10	27.15	0.12	117.05
ae	Raised floor system; steel panels 600 × 600; steel legs, bearers, feet, adjustment					
	(i) 300 high	71.01	–	15.01	0.23	86.25
	(ii) 900 high	105.22	–	21.32	0.33	126.87

Cleaning			Cyclical			Repairs					
Annual Cost	*Frequency No*	*Yr*	*Work*	*Cost*	*Cycle in Years*	*Cause/ Action*	*Cost (£) and Risk*	*Occurs in Years*	*kg/m²*	*R value*	*No*
											1
Sweep 2.38 Wash 0.72	250 50	1 1	Replace sheet, patch screed	13.24	15	Patch sheet, patch screed	22 2%	1–60	80	0.10	w
Vacuum 0.95	100	1	Replace tiles, patch screed	11.43	10	Patch tiles, patch screed	24 5%	1–60	72	0.19	x
As above 0.95	100	1	Replace carpet, patch screed	10.79	10	Patch carpet, patch screed	23 5%	1–60	73	0.17	y
Vacuum 0.95 Clean 0.09	100 1	1 5	As above	31.93	10	Relay carpet, repair carpet	47 5%	1–60	67	0.22	z
See covering	–	–	–	–	–	Damage	20 2%	1–60	15	0.36	aa
As above	–	–	–	–	–	As above	9 2%	1–60	11	0.15	ab
As above	–	–	–	–	–	Splits, dry rot, wood-worm	20 2%	1–60	13	0.18	ac
Sweep 2.38 Buff 0.72	150 50	1 1	Wax polish. Sand seal, wax polish	1.65 2.77	1 30	As above	44 1%	1–60	18	0.16	ad
											ae
–	–	–	Replace	86.28	30	Corrosion/ damage	86 1%	1–60	57	0.20	(i)
–	–	–	As above	122.52	30	As above	122 1%	1–60	70	0.20	(ii)

3.B Floor finishes—*continued*

No	Description		Capital	Cleaning	Cyclical	Repairs	Total
2	*ADD* to any above, where appropriate	m^2					
a	Expanded polystyrene, 50; fixed with adhesive		8.29	–	–	–	8.29
b	Additional cement and sand screed, 25		2.98	–	–	–	2.98
c	Impregnated fibre insulation board, 13, bedded in mastic/ bitumen		4.71	–	–	–	4.71
d	Sound deadening quilt, 13; laid loose		3.19	–	–	–	3.19
3	SKIRTINGS	m					
a	Wrought softwood, 19 × 100, rounded; prime back; prime + 2 coats front		4.86	See wall	1.99	0.02	6.87
b	Wrought hardwood, 19 × 100, rounded, Afrormosia; seal back; seal + 2 coats polyurethane clear finish front		10.26	As above	2.29	0.01	12.56
c	PVC, extruded, 100, coved, rounded; fixed with adhesive; backing		5.71	As above	2.50	0.09	8.30
d	Vinyl sheet, 2, turned up 100, over former, backing		7.13	As above	3.44	0.11	10.68
e	Quarry tile, 19 × 100, square top, cove; backing		9.77	As above	–	0.06	9.83
f	Ceramic tile, 9 × 100, rounded top, cove; backing		11.11	As above	–	0.07	11.18
g	Granolithic, 19 × 100, rounded top, cove; division strip; backing		14.83	As above	–	0.09	14.92
h	Terrazzo, 12 × 100, rounded top, cove; division strip; backing		22.30	As above	–	0.13	22.43

Cleaning			Cyclical			Repairs					
Annual Cost	Frequency No	Yr	Work	Cost	Cycle in Years	Cause/ Action	Cost (£) and Risk	Occurs in Years	kg/m²	R value	No
											2
–	–	–	–	–	–	–	–	–	2	1.67	a
–	–	–	–	–	–	–	–	–	39	0.05	b
–	–	–	–	–	–	–	–	–	4	0.23	c
–	–	–	–	–	–	–	–	–	3	0.29	d
									kg/m		3
See wall	–	–	Repaint	0.87	6	Dry rot/ woodworm	6 1%	1–60	1	–	a
As above	–	–	Re-coat	1.00	6	As above	11 0.50%	1–60	2	–	b
As above	–	–	Replace	3.76	15	Damage	7 5%	1–60	1	–	c
As above	–	–	As above	5.18	15	As above	8 5%	1–60	1	–	d
As above	–	–	–	–	–	Loosened tiles	11 2%	1–60	5	–	e
As above	–	–	–	–	–	As above	13 2%	1–60	2	–	f
As above	–	–	–	–	–	Damage	17 2%	1–60	5	–	g
As above	–	–	–	–	–	As above	24 2%	1–60	6	–	h

3.C Ceiling finishes

No	Description	Capital	Cleaning	Cyclical	Repairs	Total
1	CEILINGS m²					
a	Superfine finish to concrete	4.28	7.50	–	–	11.78
b	Emulsion paint; vinyl/matt; mist + 2 coats	2.01	6.80	3.43	0.05	12.29
c	Oil or eggshell paint; prime + 2 coats	3.32	6.80	5.35	0.08	15.55
d	Textured plastic finish, patterned	3.98	9.70	2.32	0.11	16.11
e	Plaster skim	3.62	See decoration	–	0.02	3.64
f	*ADJUST* Items 1a to 1e above where to coffered or troughed soffit — *multiply by 2.00*					
g	Plasterboard, 12; textured plastic finish; 50 × 50 sawn softwood noggings	9.39	6.80	1.93	0.03	18.15
h	Plasterboard, 12; plaster skim finish; 50 × 50 sawn softwood noggings	10.46	See decoration	–	0.04	10.50
j	Wrought softwood match boarding, 19, t&g and V jointed; two coats clear polyurethane	20.93	6.80	7.02	0.03	34.78
k	Wrought sapele match-boarding, 19, t&g and V jointed, diagonal; two coats clear polyurethane	44.77	6.80	7.02	0.06	58.65
l	Under-purlin lining; galvanised steel bearers; 12 plasterboard, plastic faced; 50 integral insulation	12.31	6.80	–	0.08	19.19
m	Suspended galvanised grid, PVC coated, exposed; wire hangers to concrete; mineral fibre tiles 600 × 600	25.45	7.50	4.99	0.08	38.02

Cleaning			*Cyclical*			*Repairs*					
Annual Cost	*Frequency No*	*Yr*	*Work*	*Cost*	*Cycle in Years*	*Cause/ Action*	*Cost (£) and Risk*	*Occurs in Years*	*kg/m²*	*R value*	*No*
											1
Dust 0.44 Wash 0.17	6 1	1 6	–	–	–	–	–	–	–	–	a
As above 0.40 0.15	6 1	1 6	Repaint 2 coats	1.50	6	Touch up	2 10%	1–60	–	–	b
As above 0.40 0.15	6 1	1 6	As above	2.34	6	As above	3 10%	1–60	–	–	c
Dust 0.60	6	1	As above	2.50	12	Touch up and repair cracks	4 10%	1–60	3	0.01	d
See dec- oration	–	–	–	–	–	Cracking, damp patches	4 1.5%	1–60	7	0.01	e
											f
Dust 0.40 Wash 0.15	6 1	1 6	Resurface	3.74	18	Joint cracks	11 1%	1–60	15	0.09	g
See dec- oration	–	–	–	–	–	Cracking, patching	13 1%	1–60	19	0.09	h
Dust 0.40 Wash 0.15	6 1	1 6	Re-coat	3.07	6	Dry rot, woodworm	24 0.50%	1–60	11	0.14	j
As above 0.40 0.15	6 1	1 6	As above	3.07	6	As above	47 0.50%	1–60	14	0.12	k
As above 0.40 0.15	6 1	1 6	–	–	–	Bowed boards	15 2%	1–60	13	1.59	l
As above 0.44 0.17	6 1	1 6	Replace completely	28.69	30	Damaged tiles	29 1%	1–60	14	0.28	m

No	3.C **Ceiling finishes** – *continued* *Description*	*Capital*	*Cleaning*	*Cyclical*	*Repairs*	*Total*
1	CEILINGS – *continued* m^2					
n	Suspended galvanised grid, PVC coated, exposed; wire hangers to concrete; 12 plasterboard panels, plastic faced, 600 × 1200	17.23	6.80	3.56	0.11	27.70
p	Suspended galvanised grid, enamelled; hangers to concrete; steel tiles, enamelled, 600 × 600	27.68	6.80	3.00	0.08	37.56
q	Suspended galvanised bearers; hangers to concrete; aluminium planks, profiled, enamelled	30.25	6.80	3.25	0.09	40.39

Cleaning			Cyclical			Repairs					
Annual Cost	Frequency No	Yr	Work	Cost	Cycle in Years	Cause/ Action	Cost (£) and Risk	Occurs in Years	kg/m²	R value	No
											1
Dust 0.40 Wash 0.15	6 1	1 6	Replace completely	20.46	30	Damaged tiles	21 2%	1–60	18	0.08	n
As above 0.40 0.15	6 1	1 6	As above	30.91	40	Damaged/ corroded tiles	31 1%	1–60	23	Ignore	p
As above 0.40 0.15	6 1	1 6	As above	33.48	40	Damaged planks	33 1%	1–60	23	Ignore	q

4 FITTINGS AND FURNISHINGS

4.A.1 Fittings, fixtures and furniture

No	Description	Capital	Cleaning	Cyclical	Repairs	Total
1	KITCHEN UNITS or similar. SET Base Cupboard 1000 × 600 × 900 (2 Nr) Sink unit base 1000 × 600 × 900 (1 Nr) Wall cupboard 1000 × 300 × 600 (2 Nr) Broom cupboard 500 × 600 × 2000 (1 Nr)					
a	Proprietary, blockboard/chipboard/plywood, painted or low cost veneer	603.90	311.10	705.01	0.18	1620.19
b	Proprietary, melamine faced chipboard, medium quality	779.71	311.10	241.45	0.23	1332.49
c	Proprietary, best quality, hardwood	1059.53	311.10	195.07	0.30	1566.00
d	Purpose made, softwood frame, plywood covering, painted.	1131.99	311.10	1049.82	0.32	2493.23
e	Purpose made, pine panelled, clear finish	1196.66	311.10	334.44	0.34	1842.54
f	Purpose made, hardwood panelled, clear finish	1928.67	311.10	308.57	0.54	2548.88
2	ENTRANCE MATS Nr					
a	Coir matting; steel mat well frame, 50 × 50 × 5 set in screed; 900 × 550	48.33	44.44	73.77	0.01	166.55
b	Rubber link mat; brass edge 50 × 5 set in screed; 900 × 550	66.74	44.44	56.43	0.02	167.63

Cleaning			Cyclical			Repairs					
Annual Cost	*Frequency No*	*Yr*	*Work*	*Cost*	*Cycle in Years*	*Cause/ Action*	*Cost (£) and Risk*	*Occurs in Years*	*kg/m²*	*R value*	*No*
											1
Wipe down 19.25	50	1	Repaint Replace	92.75 665.48	5 15	Loose joints, loose ironmongery, defective materials	666 0.10%	1–60	–	–	a
As above	50	1	Replace	841.29	25	As above	842 0.10%	1–60	–	–	b
As above	50	1	As above	1121.11	30	As above	1121 0.10%	1–60	–	–	c
As above	50	1	Repaint Replace	92.75 1184.77	5 15	As above	1185 0.10%	1–60	–	–	d
As above	50	1	Redec. Replace	50.51 1258.21	6 30	As above	1258 0.10%	1–60	–	–	e
As above	50	1	As above	50.51 1990.25	6 40	As above	1990 0.10%	1–60	–	–	f
											2
Beat and sweep 2.75	50	1	Replace mat	14.62	3	Defective materials	14 0.10%	1–60	–	–	a
As above 2.75	50	1	As above	47.18	10	As above	63 0.10%	1–60	–	–	b

5 SERVICES

5.A Sanitary appliances

No	Description	Capital	Cleaning	Cyclical	Repairs	Total
1	SANITARY FITTINGS Nr					
a	Basin, vitreous china, white, angle type, handrinse, 457 × 431; brackets; pair of taps; P trap	99.74	692.01	83.25	–	875.00
b	Basin, vitreous china, white, 560 × 405; brackets; pair of pillar taps; P trap	92.31	692.01	81.95	–	866.27
c	Basin, vitreous china, coloured, 560 × 405; brackets; pair of pillar taps; P trap	100.11	692.01	83.31	–	875.43
d	Basin, vitreous china, white, 560 × 405; pedestal; clips; pair of pillar taps; P trap	108.16	692.01	84.71	–	884.88
e	Basin, vitreous china, coloured, 560 × 405; pedestal; clips; pair of pillar taps; P trap	121.18	692.01	86.98	–	900.17
f	Sink, glazed fireclay, white, 610 × 455 × 255; brackets and legs; pair of backplate taps; P trap; teak draining board	168.39	692.01	96.84	–	957.24
g	Sink, stainless steel, 1000 × 500, single bowl, single drainer; clips to supports; pair of pillar taps; P trap	140.65	692.01	79.65	–	912.31
h	Bath, acrylic, rectangular, white/standard colour, 1700 long; pair of taps; combined trap; timber framework; plastic faced hardboard side and end panels	257.59	1153.41	80.05	–	1491.05
j	Bath, steel enamelled, white/standard colour, 1700 long; pair of taps, combined trap; timber framework; plastic faced hardboard side and end panels	278.03	1153.41	83.61	–	1515.05
k	Bath, cast iron enamelled, white/standard colour, 1700 long; pair of taps; combined trap; timber framework; enamelled hardboard side and end panels	326.48	1153.41	92.04	–	1571.93
l	WC suite, vitreous china pan, white, high level plastic cistern; plastic seat and cover	139.35	692.01	70.36	–	901.72

Cleaning			Cyclical			Repairs					
Annual Cost	*Frequency No*	*Yr*	*Work*	*Cost*	*Cycle in Years*	*Cause/ Action*	*Cost (£) and Risk*	*Occurs in Years*	*kg*	*R value*	*No*
											1
Wash 42.82	250	1	Rewasher Replace	8.11 114.17	2 30	–	–	–	14	–	a
As above 42.82	250	1	As above	8.11 106.74	2 30	–	–	–	16	–	b
As above 42.82	250	1	As above	8.11 114.54	2 30	–	–	–	16	–	c
As above 42.82	250	1	As above	8.11 122.60	2 30	–	–	–	20	–	d
As above 42.82	250	1	As above	8.11 135.61	2 30	–	–	–	20	–	e
As above 42.82	250	1	As above	8.11 192.28	2 30	–	–	–	26	–	f
As above 42.82	250	1	As above	8.11 167.77	2 30	–	–	–	5	–	g
As above 71.37	250	1	As above	8.11 289.31	4 30	–	–	–	20	–	h
As above 71.37	250	1	As above	8.11 309.75	4 30	–	–	–	30	–	j
As above 71.37	250	1	As above	8.11 358.20	4 30	–	–	–	150	–	k
As above 42.82	250	1	Rewasher Renew flush Replace	4.06 8.77 161.75	2 10 30	–	–	–	15	–	l

5.A Sanitary appliances — *continued*

No	Description		Capital	Cleaning	Cyclical	Repairs	Total
1	SANITARY FITTINGS — *continued*	Nr					
m	WC suite, vitreous china pan and cistern, high level, coloured; plastic seat and cover		189.63	692.01	79.11	–	960.75
n	WC suite, vitreous china pan, coloured, low level plastic cistern; plastic seat and cover		142.62	692.01	70.93	–	905.56
p	WC suite, vitreous china pan and cistern, low level, white; plastic seat and cover		121.90	692.01	67.33	–	881.24
q	WC suite, vitreous china pan and cistern, low level, coloured; plastic seat and cover		148.60	692.01	71.97	–	912.58
r	Urinal suite, vitreous china bowls (2Nr); cistern, white; flush pipes; traps; divider		319.70	1384.19	144.21	–	1848.10
s	Urinal suite, glazed fireclay stalls (2Nr); vitreous china cistern; flush pipes; tread; traps		822.56	2306.82	195.04	–	3324.42

Cleaning			Cyclical			Repairs					
Annual Cost	*Frequency No*	*Yr*	*Work*	*Cost*	*Cycle in Years*	*Cause/ Action*	*Cost (£) and Risk*	*Occurs in Years*	*kg*	*R value*	*No*
											1
Wash 42.82	250	1	Rewasher	4.06	2	—	—	—	18	–	m
			Renew flush	8.77	10						
			Replace	212.03	30						
As above 42.82	250	1	As above	4.06	2	—	—	—	15	–	n
				8.77	10						
				165.01	30						
As above 42.82	250	1	As above	4.06	2	—	—	—	18	–	p
				8.77	10						
				144.30	30						
As above 42.82	250	1	As above	4.06	2	—	—	—	18	–	q
				8.77	10						
				171.00	30						
As above 85.65	250	1	As above	9.56	2	—	—	—	20	–	r
				8.77	10						
				339.11	30						
As above 142.74	250	1	As above	4.06	2	—	—	—	180	–	s
				8.77	10						
				878.30	30						

5 SERVICES

5.C.1 Internal drainage

No	Description	Capital	Cleaning	Cyclical	Repairs	Total
1	DISPOSAL SYSTEM STACK AND BRANCHES					
a	MuPVC/polypropylene pipes, soil and waste 110mm × 10m + fittings; 55mm × 5m + fittings; ring seal joints	195.23	–	–	1.92	197.15
b	MuPVC pipes, soil and waste 110mm × 10m + fittings; 55mm × 5m + fittings; solvent welded joints	206.22	–	–	2.00	208.22
c	Cast iron 100mm × 10m + fittings; 50mm × 5m + fittings; Timesaver joints; painted	557.08	–	39.66	2.31	599.05
d	Cast iron 100mm × 10m + fittings; 50mm × 5m + fittings; socketed lead joints; painted	559.94	–	39.66	2.31	601.91
e	Cast iron 100mm × 10m + fittings; socketed lead joints; copper, 50mm × 5m; capilliary joints; painted	594.06	–	39.66	2.43	636.15

Cleaning			*Cyclical*			*Repairs*					
Annual Cost	*Frequency No*	*Yr*	*Work*	*Cost*	*Cycle in Years*	*Cause/ Action*	*Cost (£) and Risk*	*Occurs in Years*	*kg*	*R value*	*No*
											1
–	–	–	–	–	–	Defective joints, damage	261 1%	1–10	20	–	a
–	–	–	–	–	–	Damage	272 1%	1–10	20	–	b
–	–	–	Repaint	17.34	6	Leaking joints, obstructions	627 0.50%	1–10	220	–	c
–	–	–	As above	17.34	6	As above	627 0.50%	1–10	210	–	d
–	–	–	As above	17.34	6	As above	660 0.50%	1–10	180	–	e

5.D.2/3 Cold and hot water services

No	Description	Capital	Cleaning	Cyclical	Repairs	Total
1	DISTRIBUTION SYSTEM RISERS AND BRANCHES					
a	PVC pipework and compression fittings 19mm × 20m; 13mm × 20m	346.35	30.71	31.73	1.07	409.86
b	Galvanised steel pipework and screwed fittings 15mm × 20m; 20mm × 20m; painted	409.71	30.71	137.48	0.64	578.54
c	Copper pipework and capilliary fittings 15mm × 20m; 22mm × 20m; painted	335.09	30.71	47.26	0.27	413.33
d	Stainless steel pipework and capilliary fittings 15mm × 20m; 22mm × 20m	488.50	61.90	31.73	0.37	582.50

Cleaning			Cyclical			Repairs					
Annual Cost	*Frequency No*	*Yr*	*Work*	*Cost*	*Cycle in Years*	*Cause/ Action*	*Cost (£) and Risk*	*Occurs in Years*	*kg*	*R value*	*No*
											1
Dust 1.90	50	1	Rewasher valves	4.06	2	Damage	396 1%	1–60	7	–	a
As above 1.90	50	1	Repaint Rewasher valves	46.24 4.06	6 2	Corrosion, etc	473 0.50%	1–60	43	–	b
As above 1.90	50	1	As above	6.79 4.06	6 2	As above	407 0.25%	1–60	33	–	c
Polish 3.83	50	1	Rewasher valves	4.06	2	Damage	550 0.25%	1–60	33	–	d

5.E Heat source

No	Description	Capital	Cleaning	Cyclical	Repairs	Total
1	Heat souce to provide 1250 kw (for about 10,000 m^2 floor area); all components within plant room; water tanks, flue; insulation; builders work					
a	Water boiler, solid fuel fired; mild steel; storage space for fuel	31,350.00	5,688.67	17,074.19	44.44	54,157.30
b	Water boiler, natural gas fired, mild steel; gas main	23,320.00	5,688.67	16,116.13	34.07	45,158.87
c	Water boiler, oil fired, mild steel; oil storage tank and base	24,090.00	5,688.67	16,142.44	34.07	45,955.18
d	Steam boiler, natural gas fired, mild steel; gas main	33,660.00	5,688.67	20,205.64	47.41	59,601.73

Cleaning			Cyclical			Repairs					
Annual Cost	*Frequency No*	*Yr*	*Work*	*Cost*	*Cycle in Years*	*Cause/ Action*	*Cost (£) and Risk*	*Occurs in Years*	*kg/m²*	*R value*	*No*
											1
Degrease and wash 352.00	1	1	Checking	27.50	12/1	Replace valves, pumps, section of flue, section of boilers, leaking joints etc.	33,000 0.50%	1–60	–	–	a
			Shut down, test, clean flue and fireside; descale	557.34	1						
			Overhaul pumps, replace firebricks	292.82	2						
			Over-haul valves & traps	29.70	5						
			Repaint	110.00	10						
			Replace pumps	1,320.00	30						
As above 352.00	1	1	As above	27.50	12/1	As above	25,300 0.50%	1–60	–	–	b
				631.04	1						
				17.82	2						
				29.70	5						
				110.00	10						
				1,320.00	30						
As above 352.00	1	1	As above	27.50	12/1	As above	25,300 0.50%	1–60	–	–	c
				631.04	1						
				17.82	2						
				29.70	5						
				110.00	10						
				1,320.00	30						
As above 352.00	1	1	As above	47.30	12/1	As above	35,200 0.50%	1–60	–	–	d
				631.04	1						
				17.82	2						
				117.70	5						
				110.00	10						
				1,320.00	30						

5.E Heat source — *continued*

No	Description	Capital	Cleaning	Cyclical	Repairs	Total
1	Heat source to provide 1250 kw (for about 10,000m² floor area); all components within plant room; water tanks, flue; insulation; builders work — *continued*					
e	Steam boiler, oil fired, mild steel; storage tank and base	34,870.00	5,688.67	20,205.64	48.89	60,813.21
2	Heat pump, 78 kw; all associated equipment; builder's work					
a	Water to water	23,540.00	1,422.17	6,839.10	34.07	31,835.34
b	Air to water	14,960.00	1,422.17	6,839.10	22.22	23,243.49

Cleaning		Cyclical			Repairs					
Annual Cost	*Frequency No Yr*	*Work*	*Cost*	*Cycle in Years*	*Cause/ Action*	*Cost (£) and Risk*	*Occurs in Years*	*kg/m²*	*R value*	*No*
										1
Degrease and wash 352.00	1 1	Checking.	47.30	12/1	Replace valves, pumps, section of flue, section of boilers, leaking joints etc.	36,300 0.50%	1–60	–	–	e
		Shut down, test, clean flue and fireside; descale.	631.04	1						
		Overhaul pumps, replace firebricks.	17.82	2						
		Over-haul valves & traps.	117.70	5						
		Repaint.	110.00	10						
		Replace pumps.	1,320.00	30						
										2
Degrease and wash 88.00	1 1	Checking.	27.50	12/1	Defective parts or joints	25,300 0.50%	1–60	–	–	a
		Shut down, check, test.	62.70	1						
		Overhaul pumps.	17.82	2						
		Overhaul valves.	29.70	5						
		Repaint.	33.00	10						
		Replace pumps	1,320.00	30						
As above 88.00	1 1	As above	27.50	12/1	As above	16,500 0.50%	1–60	–	–	b
			62.70	1						
			17.82	2						
			29.70	5						
			33.00	10						
			1,320.00	30						

No	5.F Space heating & air treatment Description	Capital	Cleaning	Cyclical	Repairs	Total
1	Heat distribution (for about 10,000 m^2 floor area); builder's work					
a	Pipework (i) with radiators	151,800	15,488	48,676	222	216,186
	(ii) with sill line convectors	100,210	4,340	11,775	154	116,479
	(iii) with unit heaters	85,250	2,426	20,236	133	108,045
	(iv) with fan convectors	118,250	2,170	19,394	178	139,992

Cleaning			*Cyclical*			*Repairs*					
Annual Cost	*Frequency No*	*Yr*	*Work*	*Cost*	*Cycle in Years*	*Cause/ Action*	*Cost (£) and Risk*	*Occurs in Years*	*kg/m²*	*R value*	*No*
											1
											a
Dust 907.90	12	1	Repaint pipes	9,529.52	6	Repair leaking joints, defective valves, etc.	165,000 0.50%	1–60	–	–	(i)
Wash			Repaint rads	9,398.62	6						
50.44	2	1	Clean out	329.12	20						
			Overhaul Pumps and valves	604.12	20						
			Check valves & sensors	309.76	1						
As above 254.43	12	1	Repaint covers	2,883.54	6	As above	114,400 0.50%	1–60	–	–	(ii)
14.14	2	1	Clean out	110.00	20						
			Overhaul pumps and valves	646.80	20						
			Check valves & sensors	301.40	1						
As above 142.23	12	1	Repaint pipes	4,044.04	6	As above	99,000 0.50%	1–60	–	–	(iii)
7.90	2	1	Repaint rads	1,184.59	6						
			Clean out	110.00	20						
			Overhaul pumps and valves	768.90	20						
			Check valves & sensors	490.00	1						
As above 127.22	12	1	As above	3,423.42	6	As above	132,000 0.50%	1–60	–	–	(iv)
7.07	2	1		1,650.00	6						
				110.00	20						
				858.00	20						
				457.60	1						

5.F Space heating & air treatment — *continued*

No	Description	Capital	Cleaning	Cyclical	Repairs	Total
1	Heat distribution (for about 10,000 m^2 floor area); builder's work — *continued*					
b	Supply and extract ducts, grilles, fans					
	(i) recirculating	224,510	1,088	39,185	311	265,094
	(ii) local extract to outside	210,650	1,040	66,597	289	278,576
2	Electric heating; (for about 10,000 m^2 floor area) cable supply; control panel; builder's work					
a	Underfloor panels	154,220	–	13,205	430	167,855
b	Night storage heaters	128,590	4,019	17,441	356	150,406

Cleaning			Cyclical			Repairs					
Annual Cost	Frequency No	Yr	Work	Cost	Cycle in Years	Cause/ Action	Cost (£) and Risk	Occurs in Years	kg/m²	R value	No
											1
											b
Dust grilles 67.32	12	1	Check stats	66.00	1	Loose fixings, loose grilles, defecitve fan, defective stat	231,000 0.50%	1–60	–	–	(i)
			Check filters & dampers	247.50	1						
			Overhaul fans	440.00	1						
			Clean grilles	3,366.00	2						
			Overhaul dampers	247.50	5						
As above 64.35	12	1	As above	66.00	1	As above	214,500 0.50%	1–60	–	–	(ii)
				247.50	1						
				2,200.00	1						
				3,234.00	2						
				247.50	5						
											2
–	–	–	Check stats	99.00	1	Defective elements/ cables	159,500 1%	1–60	–	–	a
			Electrical tests	1,485.00	2						
Dust 247.50	50	1	Replace heaters.	9,900.00	30	Defective heaters	132,000 1%	1–60	–	–	b
Wash 8.25	1	6	Replace cable	1,237.50	30						

5.H.2 Electric power supplies

No	Description	Capital	Cleaning	Cyclical	Repairs	Total
1	POWER PER RING					
	Laboratory or similar – 14 twin 13A socket outlets switched, 30m girth of wall + feed to distribution board					
a	PVC cable in conduit					
	(i) plastic faced outlets; PVC conduit; chase in brick/block	739.89	–	11.50	1.04	752.43
	(ii) metal faced outlets; PVC conduit; chase in brick/block	784.76	–	11.50	1.11	797.37
	(iii) metal faced outlets; galvanised steel conduit on surface; painted	958.53	85.03	169.55	1.33	1214.44
b	PVC cable in trunking					
	(i) plastic faced outlets; PVC surface skirting/wall trunking 150 wide; buried conduit to distribution	1278.83	47.24	11.50	1.78	1339.35
	(ii) metal faced outlets; steel surface skirting/wall trunking 150 wide, stove enamelled; buried conduit to distribution	1115.09	47.24	97.72	1.63	1261.68
	(iii) metal faced outlets; aluminium surface skirting/wall trunking 150 wide, polyester powder coated; buried conduit to distribution	1200.14	47.24	97.72	1.70	1346.80

Cleaning			Cyclical			Repairs					
Annual Cost	*Frequency No*	*Yr*	*Work*	*Cost*	*Cycle in Years*	*Cause/ Action*	*Cost (£) and Risk*	*Occurs in Years*	*kg/m²*	*R value*	*No*
											1
											a
–	–	–	Replace cable	66.07	30	Defective fitting or cable	770 0.50%	1–60	–	–	(i)
–	–	–	As above	66.07	30	As above	825 0.50%	1–60	–	–	(ii)
Dust conduit 5.24 Wash conduit 0.15	50 1	1 6	Repaint conduit Replace cable	69.11 66.07	6 30	Defective fitting, cable or conduit, corrosion	990 0.50%	1–60	–	–	(ii)
											b
Dust trunking 2.87 Wash trunking 0.17	50 1	1 3	Replace cable	66.07	30	Defective fitting, cable or trunking	1320 0.50%	1–60	–	–	(i)
As above 2.87 0.17	50 1	1 3	Replace cable Paint trunking	66.07 37.70	30 6	As above	1210 0.50%	1–60	–	–	(ii)
As above 2.87 0.17	50 1	1 3	As above	66.07 37.70	30 6	As above	1265 0.50%	1–60	–	–	(iii)

5.H.3/4 Electric lighting and lighting fittings

No	Description	Capital	Cleaning	Cyclical	Repairs	Total
1	LIGHTING PER POINT					
a	Fluorescent light fitting and lamp, single 1200, plain diffuser, fixing to surface					
	(i) PVC covered cables; PVC conduit buried in brickwork/cast in concrete	100.94	26.67	27.30	0.15	155.06
	(ii) PVC covered cables; PVC conduit buried in existing work; make good plaster	164.21	26.67	27.30	0.24	218.42
	(iii) PVC covered cables; PVC trunking fixed to surface	131.05	62.54	27.30	0.19	221.08
	(iv) PVC covered cables; steel trunking fixed to surface; painted	137.02	62.54	48.02	0.19	247.77
	(v) PVC covered cables; aluminium trunking fixed to surface	208.08	62.54	37.55	0.30	308.47
	(vi) PVC covered cables; steel conduit fixed to surface; painted	134.53	62.54	48.02	0.20	245.29
	(vii) PVC covered cables; galvanised steel conduit fixed to surface; painted; weatherproof fitting	177.13	62.54	54.96	0.25	294.88
	(viii) MICC, PVC covered cable; fixed to surface	72.01	62.54	35.54	0.21	170.30

Cleaning			Cyclical			Repairs					
Annual Cost	*Frequency No*	*Yr*	*Work*	*Cost*	*Cycle in Years*	*Cause/ Action*	*Cost (£) and Risk*	*Occurs in Years*	*kg/m²*	*R value*	*No*
											1
											a
Clean diffuser 1.65	1	1	Replace lamp/ starter	3.48	4	Defective choke/ capacitor, damage to diffuser/cable	110 0.50%	0–60	–	–	(i)
			Replace fitting	31.64	20						
			Replace cable	9.24	30						
As above 1.65	1	1	As above	3.48	4	As above	176 0.50%	1–60	–	–	(ii)
				31.64	20						
				9.24	30						
Clean diffuser 1.65	1	1	As above	3.48	4	Defective choke/ capacitor, damage to diffuser/cable/ trunking	143 0.50%	1–60	–	–	(iii)
				31.64	20						
				9.24	30						
Dust trunking 2.20	50	1									
Wash trunking 0.14	1	6									
As above 1.65	1	1	As above	3.48	4	As above	143 0.50%	1–60	–	–	(iv)
2.20	50	1		31.64	20						
0.14	1	6		9.24	30						
			Repaint trunking	9.06	6						
As above 1.65	1	1	As above	3.48	4	As above	220 0.50%	1–60	–	–	(v)
2.20	50	1		31.64	20						
0.14	1	6		9.24	30						
			Degrease trunking	4.48	6						
As above 1.65	1	1	As above	3.48	4	Defective choke/ capacitor, damage to diffuser/cable/ conduit	149 0.50%	1–60	–	–	(vi)
2.20	50	1		31.64	20						
0.14	1	6		9.24	30						
			Repaint conduit	9.06	6						
As above 1.65	1	1	As above	3.48	4	As above	187 0.50%	1–60	–	–	(vii)
2.20	50	1		31.64	20						
0.14	1	6		9.24	30						
			Repaint conduit	9.75	5						
As above 1.65	1	1	As above	3.48	4	As above	77 1%	1–60	–	–	(viii)
2.20	50	1		31.64	20						
0.14	1	6		56.61	30						

5.H.3/4 Electric lighting and lighting fittings — *continued*

No	Description	Capital	Cleaning	Cyclical	Repairs	Total
1	LIGHTING — *continued* PER POINT					
b	Fluorescent light fitting and lamp, single 1200, fitted in lay-in proprietary suspended ceiling; perspex panel 600 × 1200; excluding suspension and framework; PVC covered cable fixed to surface	61.56	19.55	28.87	0.09	110.07
c	Tungsten lamp in pendant with plain glass shade; flex; rose; PVC covered cables; PVC conduit buried in brickwork/cast in concrete	94.80	17.78	25.89	0.13	138.60
d	Tungsten lamp; 380mm bowl fitting, PVC covered cables, PVC conduit buried in brickwork/cast in concrete	93.68	17.78	37.79	0.14	149.39
e	Tungsten lamp 2D type; 380mm bowl fitting, PVC covered cables, PVC conduit buried in brickwork/cast in concrete	98.56	17.78	39.38	0.14	155.86

Cleaning			Cyclical			Repairs					
Annual Cost	*Frequency No*	*Yr*	*Work*	*Cost*	*Cycle in Years*	*Cause/ Action*	*Cost (£) and Risk*	*Occurs in Years*	*kg/m²*	*R value*	*No*
											1
Clean perspex 1.21	1	1	Replace lamp/ starter Replace perspex Replace cable	3.48 9.71 69.80	4 20 30	Defective choke/ capacitor, damage to diffuser/cable	66 0.50%	1–60	–	–	b
Clean shade 1.10	1	1	Replace lamp Replace shade Replace fitting Replace cable	1.28 6.53 15.86 9.14	2 10 20 30	As above	99 0.50%	1–60	–	–	c
Clean bowl 1.10	1	1	Replace lamp Replace fitting Replace cable	1.28 21.90 9.14	2 10 30	As above	105 0.50%	1–60	–	–	d
As above 1.10	1	1	As above	6.07 21.90 9.14	7 10 30	As above	105 0.50%	1–60	–	–	e

5.J.1 Lifts

No	Description	Capital	Cleaning	Cyclical	Repairs	Total
1	Two × 10 person passenger lifts; 6 storey; doors and architraves; controls; builder's work; excluding cost of lift shaft.					
a	Electric traction type; overhead motor room not included.					
	(i) standard finishes and controls	140,415	12,800	66,994	46,220	266,429
	(ii) best quality finishes and controls	187,853	12,800	85,320	46,220	332,193
b	Electric hydraulic type; pit motor room not included					
	(i) standard finishes and controls	139,150	12,800	66,994	46,220	265,164
	(ii) best quality finishes and controls	179,630	12,800	82,001	46,220	320,651
c	Prestige open glazed, blister type; motor room not included	271,975	25,599	117,297	47,998	462,869

Cleaning			Cyclical			Repairs					
Annual Cost	Frequency No	Yr	Work	Cost	Cycle in Years	Cause/ Action	Cost (£)	Occurs in Years	t	R value	No
											1
											a
Wipe down 792	250	1	Repaint	1,056	5#	Maintenance agreement	2,860	1–60	2	–	(i)
			Replace floor coverings	264	5						
			Replace all	158,015	20						
As above 792	250	1	Replace floor coverings	352	4	As above	2,860	1–60	2	–	(ii)
			Replace all	205,453	20						
											b
As above 792	250	1	Repaint	1,056	5#	As above	2,860	1–60	2	–	(i)
			Replace floor covering	264	5						
			Replace all	158,015	20						
As above 792	250	1	Replace floor coverings	352	4	As above	2,860	1–60	2	–	(ii)
			Replace all	197,339	20						
Wash 1,584	250	1	As above	550	5	As above	2,970	1–60	2	–	c
				282,975	20						

Commencing year 15

6.A.2 Surface treatment

No	Description		Capital	Cleaning	Cyclical	Repairs	Total
1	LANDSCAPING	m²					
a	Cultivating surface of ground, one spit deep, remove debris, weeding		0.99	6.95	23.76	–	31.70
b	Seeding, grass seed (PC £2.60/kg); 0.042kg/m²; applied twice		0.66	7.60	29.09	0.05	37.40
c	Turfing, selected meadow turf (PC £0.80/m²); rolling		2.70	7.60	29.09	0.05	39.44
2	FOOTPATHS	m²					
a	Gravel, 50; hardcore 150; excavation		8.84	8.89	0.76	0.29	18.78
b	Bredon gravel, base 50, top layer 6; hardcore 150; excavation		12.77	8.89	0.76	0.41	22.83
c	Concrete, plain in situ, 100, surface ribbed or similar; hardcore 150; excavation		19.58	8.89	0.02	0.30	28.79
d	Concrete, reinforced in situ, 100, surface ribbed or similar; fabric reinforcement, 2.61 kg/m²; hardcore 150; excavation		21.98	8.89	0.02	0.35	31.24
e	Cobblestone paving; concrete in situ bed 100: hardcore 150: excavation		62.85	8.89	6.17	0.89	78.80
f	Asphalt 15; chippings; tack coat; concrete, reinforced in situ, 100; fabric reinforcement, 2.61 kg/m²; hardcore 150; excavation		26.25	8.89	2.76	0.55	38.45

Cleaning			Cyclical			Repairs					
Annual Cost	*Frequency No*	*Yr*	*Work*	*Cost*	*Cycle in Years*	*Cause/ Action*	*Cost (£) and Risk*	*Occurs in Years*	*kg/m²*	*R value*	*No*
											1
Hand weed/fork/ hoe 0.43	2	1	Recultivating Weedkiller	0.21 0.28	3/1 3/1	–	–	–	–	–	a
Remove litter 0.44 Remove leaves 0.03	12 2	1 1	Grass cutting Fertilizer	0.20 0.30	6/1 2/1	Repair worn areas	1.36 15%	1–60	–	–	b
As above 0.44 0.03	12 2	1 1	As above	0.20 0.30	6/1 2/1	As above	1.36 15%	1–60	–	–	c
											2
Manual sweep 0.55	50	1	Rake and roll	0.15	3	Scarify, rake, add material, roll	3.58 30%	1–60	–	–	a
As above 0.55	50	1	As above	0.15	3	Scarify, rake, add fine material, roll	5.06 30%	1–60	–	–	b
As above 0.55	50	1	Treat moss	0.12	30	Defective work/ materials	22 5%	1–60	–	–	c
As above 0.55	50	1	As above	0.12	30	As above	26 5%	1–60	–	–	d
As above 0.55	50	1	Re-bed cobbles	35.45	30	Dislodged cobbles	66 5%	1–60	–	–	e
As above 0.55	50	1	Renew asphalt	6.74	20	Isolated defects	29 7%	1.60	–	–	f

6.A.2 Surface treatment — *continued*

No	Description	Capital	Cleaning	Cyclical	Repairs	Total
2	FOOTPATHS — *continued* m²					
g	Macadam, base course 38, wearing course 12, chippings; hardcore 150; excavation	19.82	8.89	8.60	0.41	37.72
h	Precast concrete slabs 50, natural colour, jointed, bedded 50 cement and sand, hardcore 150; excavation	19.54	8.89	2.52	0.59	31.54
j	Precast concrete slabs 50, coloured, jointed, bedded 50 cement and sand, hardcore 150; excavation	20.91	8.89	2.76	0.67	33.23
k	Precast concrete slabs 50, natural colour, jointed, bedded 25 cement and sand; concrete plain in situ 100; hardcore 150; excavation	31.85	8.89	–	0.68	41.42
l	Precast concrete slabs 50, coloured, jointed, bedded 25 cement and sand; concrete plain in situ 100; hardcore 150; excavation	33.21	8.89	–	0.70	42.80
m	Artificial stone 50, jointed, bedded 50 cement and sand; hardcore 150; excavation	24.67	8.89	3.41	0.75	37.72
n	Artificial stone 50, jointed, bedded 25 cement and sand; concrete plain in situ 100; hardcore 150; excavation	36.97	8.89	–	0.75	46.61
p	Natural stone 75, jointed, bedded 50 cement and sand; hardcore 150; excavation	133.19	8.89	22.29	1.86	166.23
q	Natural stone 75, jointed, bedded 25 cement and sand; concrete plain in situ 100; hardcore 150; excavation	145.57	8.89	–	0.80	155.26
r	Perforated concrete blocks 110, interlocking; sand bed 50; granular bed 150, excavation; topsoil infill; fertilizer; grass seed	34.25	2.10	2.91	1.05	40.31
s	Granite setts 125, jointed, bedded 25 cement and sand; concrete plain in situ 100; hardcore 150; excavation	78.74	8.89	0.52	0.22	88.37
t	Brick paviors 50, clay (PC £310.00/K), laid flat, jointing cement and sand; bedding sand 50; compacted; hardcore 150, blinding; excavation	41.21	8.89	8.72	1.19	60.01

Cleaning			Cyclical			Repairs					
Annual Cost	*Frequency No*	*Yr*	*Work*	*Cost*	*Cycle in Years*	*Cause/ Action*	*Cost (£) and Risk*	*Occurs in Years*	*kg/m²*	*R value*	*No*
											2
Manual sweep 0.55	50	1	Resurface macadam	21.02	20	Isolated defects	22 7%	1–60	–	–	g
As above 0.55	50	1	Renew slabs	14.48	30	Loosened/ damaged slabs	22 10%	1–60	–	–	h
As above 0.55	50	1	As above	15.84	30	As above	25 10%	1–60	–	–	j
As above 0.55	50	1	–	–	–	As above	36 7%	1–60	–	–	k
As above 0.55	50	1	–	–	–	As above	37 7%	1–60	–	–	l
As above 0.55	50	1	Replace slabs	19.62	30	As above	28 10%	1–60	–	–	m
As above 0.55	50	1	–	–	–	As above	40 7%	1–60	–	–	n
As above 0.55	50	1	Replace slabs	128.13	30	As above	149 5%	1–60	–	–	p
As above 0.55	50	1	–	–	–	As above	149 2%	1–60	–	–	q
As above 0.13	12	1	Grass cutting	0.03	6/1	Re-lay blocks and base	39 10%	1–60	–	–	r
As above 0.55	50	1	Repoint	3.01	30	Re-bed loose setts	83 1%	1–60	–	–	s
As above 0.55	50	1	Repoint Re-lay and re-bed	4.43 19.66	10 30	Re-bed loose bricks	44 10%	1–60	–	–	t

6.A.2 Surface treatment – *continued*

No	Description		Capital	Cleaning	Cyclical	Repairs	Total
2	FOOTPATHS – *continued*	m^2					
u	Brick paviors 50, clay (PC £310.00/K), laid flat, jointing cement and sand; bedding sand 50; compacted; herringbone pattern; hardcore 150, blinding; excavation		44.47	8.89	9.15	1.27	63.78
v	Brick paviors 50, clay (PC £310.00/K), laid flat, jointing cement and sand; bedding sand 50; compacted; concrete in situ 100; hardcore 150; excavation		52.12	8.89	5.30	1.04	67.35
w	Brick paviors 50, clay (PC £310.00/K), laid flat, jointing cement and sand; bedding sand 50; compacted; herringbone pattern; concrete in situ 100; hardcore 150; excavation		55.39	8.89	5.30	1.09	70.67
x	Concrete blocks, rectangular 200 × 100 × 65 (PC £6.00/m^2); jointing cement and sand; bedding sand 50; compacting; hardcore 150; blinding; excavation		33.58	8.89	4.48	0.97	47.92
y	Concrete blocks, rectangular 200 × 100 × 65 (PC £6.00/m^2); jointing cement and sand; bedding sand 50; compacting; herringbone pattern; hardcore 150; blinding; excavation		36.83	8.89	4.91	1.08	51.71
z	Concrete blocks, interlocking 200 × 100 × 60 (PC £8.00/m^2); jointing cement and sand; bedding sand 50; compacting; hardcore 150; blinding; excavation		36.67	8.89	4.48	1.05	51.09
3	VEHICLE WAYS	m^2					
a	Gravel 75; hardcore 225; excavation		13.11	8.89	0.76	0.29	23.05
b	Bredon gravel, base 75, top layer 6; hardcore 225; excavation		18.61	8.89	0.76	0.41	28.67
c	Concrete, plain in situ 150, bays 25 m^2, surface tamped or similar; contraction joints; sand blinding 25; granular base, DOT type 1225; excavation		32.11	4.53	1.22	0.50	38.36
d	Concrete, reinforced in situ 150, bays 25 m^2, surface tamped or similar; fabric reinforcment 2.61 kg/m^2; expansion joints; polythene; sand blinding 25; granular base, DOT type 1225; excavation		35.42	4.53	1.22	0.35	41.52

Cleaning			Cyclical			Repairs					
Annual Cost	*Frequency No*	*Yr*	*Work*	*Cost*	*Cycle in Years*	*Cause/ Action*	*Cost (£) and Risk*	*Occurs in Years*	*kg/m²*	*R value*	*No*
											2
Manual sweep 0.55	50	1	Repoint Re-lay and re-bed	4.43 22.14	10 30	Re-lay loose bricks	47 10%	1–60	–	–	u
As above 0.55	50	1	Repoint	4.43	10	As above	55 7%	1–60	–	–	v
As above 0.55	50	1	As above	4.43	10	As above	58 7%	1–60	–	–	w
As above 0.55	50	1	Re-lay and re-bed	25.75	30	Re-bed loose blocks	36 10%	1–60	–	–	x
As above 0.55	50	1	As above	28.24	30	As above	40 10%	1–60	–	–	y
As above 0.55	50	1	As above	25.75	30	As above	39 10%	1–60	–	–	z
											3
Manual sweep 0.55	50	1	Rake and roll	0.15	3	Scarify, rake, add material, roll	3.58 30%	1–60	–	–	a
As above 0.55	50	1	As above	0.15	3	As above	5.06 30%	1–60	–	–	b
Mechanical sweep 0.28	50	1	Resurface with asphalt	5.36	30 and 50	Defective work/ materials	37 5%	1–60	–	–	c
As above 0.28	50	1	As above	5.36	30 and 50	As above	43 3%	1–60	–	–	d

6.A.2 Surface treatment — *continued*

No	Description	Capital	Cleaning	Cyclical	Repairs	Total
3	VEHICLE WAYS — *continued* m²					
e	Concrete, reinforced in situ 175, bays 25 m², surface tamped or similar; fabric reinforcment 2.61 kg/m²; expansion joints; polythene; sand blinding 50; granular base, DOT type 1230; excavation	38.38	4.53	1.22	0.37	44.50
f	Asphalt 19; chippings; tack coat; concrete reinforced in situ 150; fabric reinforcement 2.61 kg/m²; polythene; sand blinding 25; granular base, DOT type 1225; excavation	40.24	4.53	2.19	0.83	47.79
g	Macadam, base course 50, wearing course 25, chippings; tack coat; concrete weak in situ 225; granular base, DOT type 1150; excavation	49.21	4.53	2.67	1.04	57.45
h	Perforated concrete blocks 125, interlocking, vibrated; sand bed 50; hardcore 150; granular sub base, DOT type 1150, excavation; topsoil infill; fertilizer; grass seed	42.93	8.89	2.91	1.24	55.97
j	Granite setts 125, jointed, bedded 25, cement and sand; concrete weak in situ 150; granular base, DOT type 1150; excavation	93.36	4.53	0.52	0.27	98.68
k	Brick paviors 102, clay (PC £330.00/K), laid on edge, jointing cement and sand; bedding sand 50; compacted; granular base, DOT type 1150; excavation	67.06	4.53	15.80	1.94	89.33
l	Brick paviors 102, clay (PC £330.00/K), laid on edge; herringbone pattern, jointing cement and sand; bedding sand 50, compacted; granular base, DOT type 1225; excavation	72.56	4.53	15.80	2.07	94.96
m	Brick paviors 102, clay (PC £330.00/K), laid on edge; jointing cement and sand; bedding sand 50, compacted; concrete reinforced in-situ 150; fabric reinforcement 2.61 kg/m²; granular base, DOT type 1225; excavation	87.21	4.53	15.80	1.73	109.27
n	Brick paviors 102, clay (PC £330.00/K), laid on edge, herringbone pattern, jointing cement and sand; bedding sand 50, compacted; concrete reinforced in-situ 150; fabric reinforcement 2.61 kg/m²; granular base, DOT type 1225; excavation	92.72	4.53	15.80	1.85	114.90

Cleaning			Cyclical			Repairs					
Annual Cost	*Frequency No*	*Yr*	*Work*	*Cost*	*Cycle in Years*	*Cause/ Action*	*Cost (£) and Risk*	*Occurs in Years*	*kg/m²*	*R value*	*No*
											3
Mechanical sweep 0.28	50	1	Resurface with asphalt	5.36	30 and 50	Defective work/ materials	46 3%	1–60	–	–	e
As above 0.28	50	1	Re-surface asphalt	5.36	20	Defective areas	44 7%	1–60	–	–	f
As above 0.28	50	1	Resurface macadam	6.53	20	As above	55 7%	1–60	–	–	g
Manual brush 0.55	50	1	Grass cutting	0.03	6/1	Re-lay blocks and base	46 10%	1–60	–	–	h
Mechanical sweep 0.28	50	1	Repointing	3.01	30	Defective isolated areas	99 1%	1–60	–	–	j
As above 0.28	50	1	Repointing Replace bricks	5.09 55.83	10 30	As above	72 10%	1–60	–	–	k
As above 0.28	50	1	As above	5.09 55.83	10 30	As above	77 10%	1–60	–	–	l
As above 0.28	50	1	As above	5.09 55.83	10 30	As above	92 7%	1–60	–	–	m
As above 0.28	50	1	As above	5.09 55.83	10 30	As above	98 7%	1–60	–	–	n

6.A.2 Surface treatment — *continued*

No	Description		Capital	Cleaning	Cyclical	Repairs	Total
3	VEHICLE WAYS — *continued*	m^2					
p	Concrete blocks, rectangular 200 × 100 × 80 (PC £7.00/m^2), jointing cement and sand; bedding sand 50; compacting, granular base, DOT type 1225; excavation		40.48	4.53	7.85	1.19	54.05
q	Concrete blocks, rectangular 200 × 100 × 80 (PC £7.00/m^2), jointing cement and sand; bedding sand 50; herringbone pattern; compacting; concrete weak in situ 225; granular base, DOT type 1225; excavation		62.44	4.53	2.18	1.78	70.93
r	Concrete blocks, interlocking 200 × 100 × 80 (PC £8.80/m^2), jointing cement and sand; bedding sand 50; compacting; granular base, DOT type 1225; excavation		42.98	4.53	7.85	1.27	56.63

Cleaning		*Cyclical*			*Repairs*					
Annual Cost	*Frequency No Yr*	*Work*	*Cost*	*Cycle in Years*	*Cause/ Action*	*Cost (£) and Risk*	*Occurs in Years*	*kg/m²*	*R value*	*No*
										3
Mechanical sweep 0.28	50 1	Re-lay and re-bed blocks	19.20	20	Replace small areas	44 10%	1–60	–	–	p
As above 0.28	50 1	As above	22.44	40	As above	66 10%	1–60	–	–	q
As above 0.28	50 1	As above	19.20	20	As above	47 10%	1–60	–	–	r

6.A.2 Surface treatment — *continued*

No	*Description*	*Capital*	*Cleaning*	*Cyclical*	*Repairs*	*Total*
4	EDGES 10 m					
a	Precast concrete kerb 127 × 254, bedded, jointed cement and sand; concrete in situ foundation; excavation	197.90	–	–	4.15	202.05
b	Granite kerb 150 × 300, bedded cement and sand, dry jointed; concrete in situ foundation; excavation	590.54	–	–	3.30	593.84
c	Engineering brick on edge kerb 215 × 102, bedded, jointed cement and sand; concrete in situ foundation; excavation	325.02	–	10.48	4.69	340.19
d	Precast concrete edging 51 × 152, bedded jointed cement and sand; concrete in situ foundation; excavation	138.41	–	–	2.83	141.24
e	Sawn softwood edging 25 × 100; posts 50 × 50 × 400, 2m centres; treated	40.57	–	50.76	1.16	92.49

Cleaning		Cyclical			Repairs					
Annual Cost	*Frequency No Yr*	*Work*	*Cost*	*Cycle in Years*	*Cause/ Action*	*Cost (£) and Risk*	*Occurs in Years*	*kg/m²*	*R value*	*No*
										4
–	– –	–	–	–	Isolated defects	220 7%	1–60	–	–	a
–	– –	–	–	–	As above	613 2%	1–60	–	–	b
–	– –	Repointing	25.63	20	As above	348 5%	1–60	–	–	c
–	– –	–	–	–	As above	150 7%	1–60	–	–	d
–	– –	Replacing	42.44	10	As above	43 10%	1–60	–	–	e

6.A.3 Site enclosure and division

No	Description	Capital	Cleaning	Cyclical	Repairs	Total
1	BOUNDARY WALLS 10m					
a	Wall, half brick, facings (PC £300.00/K), pointed both sides; 900 high; piers, 2m centres; brick on end coping; damp proof course; concrete in situ foundation; excavation	935.92	–	2.53	0.45	938.90
b	Wall, one brick, facings (PC £300.00/K), pointed both sides; 1800 high; brick on edge coping; damp proof course; concrete in situ foundation; excavation	2419.69	–	2.53	1.16	2423.38
2	FENCES 10m					
a	Concrete posts, 3.5m centres; galvanised wires, strained; concrete in situ bases					
	(i) 850 high, 3 lines	67.13	–	1.73	0.97	69.83
	(ii) 1000 high, 5 lines	95.78	–	2.88	1.33	99.99
b	Steel angle posts, 3m centres; galvanised chain link mesh; galvanised line wires; concrete in situ bases					
	(i) 900 high, 2 lines	90.45	–	24.43	1.27	116.15
	(ii) 1800 high, 3 lines	161.23	–	40.45	2.22	203.90
c	Concrete posts, 3m centres; galvanised chain link mesh; galvanised line wires; concrete in situ bases					
	(i) 900 high, 2 lines	104.92	–	24.43	1.48	130.83
	(ii) 1800 high, 3 lines	168.74	–	40.45	2.34	211.53
d	Steel angle posts, 3m centres; plastic coated chain link mesh; plastic coated line wires; concrete in situ bases					
	(i) 900 high, 2 lines	101.82	–	6.36	1.41	109.59
	(ii) 1800 high, 3 lines	165.98	–	9.72	2.32	178.02

Cleaning			Cyclical			Repairs					
Annual Cost	*Frequency No*	*Yr*	*Work*	*Cost*	*Cycle in Years*	*Cause/ Action*	*Cost (£) and Risk*	*Occurs in Years*	*kg/m²*	*R value*	*No*
											1
–	–	–	Repointing	14.55	30	Cracks	990 0.25%	10–60	–	–	a
–	–	–	As above	14.55	30	As above	2530 0.25%	10–60	–	–	b
											2
											a
–	–	–	Replace lines	9.93	30	Replace posts	72 5%	1–60	–	–	(i)
–	–	–	As above	16.54	30	As above	99 5%	1–60	–	–	(ii)
											b
–	–	–	Replace mesh	59.72	20	As above	94 5%	1–60	–	–	(i)
–	–	–	As above	98.90	20	As above	165 5%	1–60	–	–	(ii)
											c
–	–	–	As above	59.72	20	As above	110 5%	1–60	–	–	(i)
–	–	–	As above	98.90	20	As above	174 5%	1–60	–	–	(ii)
											d
–	–	–	As above	65.56	40	As above	105 5%	1–60	–	–	(i)
–	–	–	As above	100.16	40	As above	172 5%	1–60	–	–	(ii)

6.A.3 Site enclosure and division — *continued*

No	*Description*	*Capital*	*Cleaning*	*Cyclical*	*Repairs*	*Total*
2	FENCES — *continued* 10m					
e	Concrete posts 3m centres; plastic coated chain link mesh; plastic coated line wires; concrete in situ bases					
	(i) 900 high, 2 lines	107.26	–	6.36	1.51	115.13
	(ii) 1800 high, 3 lines	173.97	–	9.72	2.41	186.10
f	Cleft chestnut pales; sweet chestnut posts, driven					
	(i) 1050 high, posts 2.5 m centres	66.88	–	16.73	2.59	86.20
	(ii) 1350 high, posts 2.25m centres	88.21	–	22.71	3.53	114.45
g	Close boarded, softwood pales, rails and capping, preservative treated; concrete posts 3m centres; concrete gravel board; in situ concrete bases					
	(i) 1200 high	246.77	–	110.99	7.41	365.17
	(ii) 1800 high	348.49	–	166.41	10.07	524.97
h	Close boarded, oak pales, rails, cappings and posts, preservative treated; posts 3m centres; oak gravel board; in situ concrete bases					
	(i) 1200 high	283.73	–	52.38	9.48	345.59
	(ii) 1800 high	396.59	–	73.80	11.85	482.24
j	Palisade, oak pales, rails and posts, preservative treated; posts 3m centres; concrete in situ bases					
	(i) 1200 high	312.47	–	46.85	8.89	368.21
	(ii) 1800 high	472.68	–	76.06	13.33	562.07

Cleaning			Cyclical			Repairs					
Annual Cost	Frequency No	Yr	Work	Cost	Cycle in Years	Cause/ Action	Cost (£) and Risk	Occurs in Years	kg/m²	R value	No
											2
											e
–	–	–	Replace mesh	65.56	40	Replace posts	112 5%	1–60	–	–	(i)
–	–	–	As above	100.16	40	As above	179 5%	1–60	–	–	(ii)
											f
–	–	–	Replace fence	96.17	30	Resecure fence	96 10%	1–60	–	–	(i)
–	–	–	As above	130.50	30	As above	131 10%	1–60	–	–	(ii)
											g
–	–	–	Replace all softwood. Re-treat with preservative	293.11 50.16	30 10	Replace posts	275 10%	1–60	–	–	(i)
–	–	–	As above	439.23 75.24	30 10	As above	374 10%	1–60	–	–	(ii)
											h
–	–	–	Replace all oak. Re-treat with preservative	353.20 50.16	50 15	As above	352 10%	1–60	–	–	(i)
–	–	–	As above	441.53 75.24	50 15	As above	440 10%	1–60	–	–	(ii)
											j
			As above	221.45 52.55	50 15	As above	330 10%	1–60	–	–	(i)
			As above	439.23 78.83	50 15	As above	495 10%	1–60	–	–	(ii)

6.A.3 Site enclosure and division — *continued*

No	Description	Capital	Cleaning	Cyclical	Repairs	Total
2	FENCES — *continued* 10m					
k	Woven softwood panels, softwood capping, oak posts, preservative treated; posts 1.8m centres; concrete in situ bases					
	(i) 1200 high	136.62	–	119.15	7.70	263.47
	(ii) 1800 high	172.46	–	155.07	9.48	337.01
l	Precast concrete panels 40; precast concrete posts 1.8m centres; concrete in situ bases; 1800 high	565.04	–	–	4.62	569.66
m	Steel security palisade type or similar; posts 2.5m centres; concrete in situ bases; painted; 2400 high	646.53	–	265.73	1.78	914.04
n	*ADD* where grounds not maintained – for clearing 1m both sides of fence of bushes, scrub, undergrowth, grass, weeds, saplings under 150 girth	–	309.64	–	–	309.64

Cleaning			Cyclical			Repairs					
Annual Cost	*Frequency No*	*Yr*	*Work*	*Cost*	*Cycle in Years*	*Cause/ Action*	*Cost (£) and Risk*	*Occurs in Years*	*kg/m²*	*R value*	*No*
											2
											k
–	–	–	Replace panels Re-treat with preservative	144.63 50.16	20 10	Replace posts	143 20%	1–60	–	–	(i)
–	–	–	As above	159.12 75.24	20 10	As above	176 20%	1–60	–	–	(ii)
–	–	–	–	–	–	Isolated replacements	572 3%	1–60	–	–	l
–	–	–	Repaint Replace sections	70.47 6.47	6 1	Isolated repairs	660 1%	1–60	–	–	m
Clear 19.16	1	1	–	–	–	–	–	–	–	–	n

6.B Drainage

No	Description	Capital	Cleaning	Cyclical	Repairs	Total
1	PIPEWORK 10m					
a	UPVC pipes and fittings, ring seal joints; granular bed and surround; excavation					
	(i) 100 nominal; 750 deep	198.54	–	–	0.59	199.13
	(ii) 150 nominal; 750 deep	267.86	–	–	0.77	268.63
b	UPVC pipes and fittings, ring seal joints; concrete bed and haunching; excavation					
	(i) 100 nominal; 750 deep	218.21	–	–	0.65	218.86
	(ii) 150 nominal; 750 deep	297.88	–	–	0.89	298.77
c	UPVC pipes and fittings, ring seal joints; concrete bed and surround; excavation; 150 nominal; 750 deep	368.87	–	–	1.07	369.94
d	Vitrified clay pipes and fittings, super strength, push-fit polypropylene couplings; granular bed and surround; excavation					
	(i) 100 nominal; 750 deep	192.87	–	–	0.56	193.43
	(ii) 150 nominal; 750 deep	246.05	–	–	0.71	246.76
e	Vitrified clay pipes and fittings, super strength, push-fit polypropylene couplings; concrete bed and haunching; excavation					
	(i) 100 nominal; 750 deep	216.55	–	–	0.65	217.20
	(ii) 150 nominal; 750 deep	329.01	–	–	0.98	329.99

Cleaning			Cyclical			Repairs					
Annual Cost	Frequency No	Yr	Work	Cost	Cycle in Years	Cause/ Action	Cost (£) and Risk	Occurs in Years	kg/m^2	R value	No
											1
											a
–	–	–	–	–	–	Obstructions, subsidence, crushing, cracking, root entry	220 1%	1–60	–	–	(i)
–	–	–	–	–	–	As above	286 1%	1–60	–	–	(ii)
											b
–	–	–	–	–	–	As above	242 1%	1–60	–	–	(i)
–	–	–	–	–	–	As above	330 1%	1–60	–	–	(ii)
–	–	–	–	–	–	As above	396 1%	1–60	–	–	c
											d
–	–	–	–	–	–	As above	209 1%	1–60	–	–	(i)
–	–	–	–	–	–	As above	264 1%	1–60	–	–	(ii)
											e
–	–	–	–	–	–	As above	242 1%	1–60	–	–	(i)
–	–	–	–	–	–	As above	363 1%	1–60	–	–	(ii)

6.B Drainage — *continued*

No	Description	Capital	Cleaning	Cyclical	Repairs	Total
1	PIPEWORK — *continued* 10m					
f	Vitrified clay pipes and fittings, extra strength; push-fit socketted joints; granular bed and surround; excavation					
	(i) 100 nominal; 750 deep	210.30	–	–	0.62	210.92
	(ii) 150 nominal; 750 deep	255.93	–	–	0.74	256.67
	(iii) 225 nominal; 1000 deep	437.40	–	–	1.24	438.64
	(iv) 300 nominal; 1500 deep	663.05	–	–	1.84	664.89
g	Vitrified clay pipes and fittings, extra strength; push-fit socketted joints; concrete bed and haunching; excavation					
	(i) 100 nominal; 750 deep	229.97	–	–	0.34	230.31
	(ii) 150 nominal; 750 deep	285.96	–	–	0.41	286.37
	(iii) 225 nominal; 1000 deep	537.41	–	–	0.76	538.17
h	Vitrified clay pipes and fittings, super strength; push-fit polypropylene couplings; concrete bed and surround; excavation; 100 nominal; 750 deep	274.01	–	–	0.40	274.41
j	Vitrified clay pipes and fittings, tested; cement and sand joints; granular bed and surround; excavation					
	(i) 150 nominal; 750 deep	240.59	–	–	0.36	240.95
	(ii) 225 nominal; 1000 deep	411.38	–	–	0.58	411.96
	(iii) 300 nominal; 1500 deep	621.31	–	–	0.87	622.18
k	Vitrified clay pipes and fittings, tested; cement and sand joints; concrete bed and haunching; excavation					
	(i) 100 nominal; 750 deep	220.10	–	–	0.37	220.47
	(ii) 150 nominal; 750 deep	267.72	–	–	0.39	268.11
	(iii) 225 nominal; 1000 deep	507.66	–	–	0.71	508.37

Cleaning			Cyclical			Repairs					
Annual Cost	Frequency No	Yr	Work	Cost	Cycle in Years	Cause/ Action	Cost (£) and Risk	Occurs in Years	kg/m²	R value	No
											1
											f
–	–	–	–	–	–	Obstructions, subsidence, crushing, cracking, root entry	231 1%	1–60	–	–	(i)
–	–	–	–	–	–	As above	275 1%	1–60	–	–	(ii)
–	–	–	–	–	–	As above	462 1%	1–60	–	–	(iii)
–	–	–	–	–	–	As above	682 1%	1–60	–	–	(iv)
											g
–	–	–	–	–	–	As above	253 0.50%	1–60	–	–	(i)
–	–	–	–	–	–	As above	308 0.50%	1–60	–	–	(ii)
–	–	–	–	–	–	As above	561 0.50%	1–60	–	–	(iii)
–	–	–	–	–	–	As above	297 0.50%	1–60	–	–	h
											j
–	–	–	–	–	–	As above	264 0.50%	1–60	–	–	(i)
–	–	–	–	–	–	As above	429 0.50%	1–60	–	–	(ii)
–	–	–	–	–	–	As above	649 0.50%	1–60	–	–	(iii)
											k
–	–	–	–	–	–	As above	275 0.50%	1–60	–	–	(i)
–	–	–	–	–	–	As above	286 0.50%	1–60	–	–	(ii)
–	–	–	–	–	–	As above	528 0.50%	1–60	–	–	(iii)

6.B Drainage — *continued*

No	Description	Capital	Cleaning	Cyclical	Repairs	Total
1	PIPEWORK — *continued* 10m					
l	Vitrified clay pipes and fittings, cement and sand joints; concrete bed and surround; excavation 100 nominal; 750 deep	341.61	–	–	0.51	342.12
m	Concrete pipes and fittings; flexible joints; granular bed and surround; excavation					
	(i) 100 nominal; 750 deep	275.75	–	–	0.40	276.15
	(ii) 225 nominal; 1000 deep	422.51	–	–	0.59	423.10
	(iii) 300 nominal; 1500 deep	580.36	–	–	0.81	581.17
n	Cast iron pipes and fittings; bolted flexible couplings; granular bed and surround; excavation					
	(i) 100 nominal; 750 deep	480.60	–	–	0.34	480.94
	(ii) 150 nominal; 750 deep	726.31	–	–	0.50	726.81
p	Cast iron pipes and fittings; bolted flexible couplings; concrete bed and haunching; flexible joints in concrete; excavation					
	(i) 100 nominal; 750 deep	500.35	–	–	0.35	500.70
	(ii) 150 nominal; 750 deep	756.33	–	–	0.53	756.86
q	Cast iron pipes and fittings; bolted flexible couplings; concrete bed and surround; flexible joints in concrete; excavation; 100 nominal; 750 deep	827.32	–	–	0.57	827.89
r	Cast iron pipes and fittings; lead socketted joints; granular bed and surround; excavation					
	(i) 100 nominal; 750 deep	853.93	–	–	0.59	854.52
	(ii) 225 nominal; 1000 deep	2,326.59	–	–	1.59	2,328.18

Cleaning			*Cyclical*			*Repairs*					
Annual Cost	*Frequency No*	*Yr*	*Work*	*Cost*	*Cycle in Years*	*Cause/ Action*	*Cost (£) and Risk*	*Occurs in Years*	*kg/m²*	*R value*	*No*
											1
–	–	–	–	–	–	Obstructions, subsidence, crushing, cracking, root entry	376 0.50%	1–60	–	–	l
											m
–	–	–	–	–	–	As above	297 0.50%	1–60	–	–	(i)
–	–	–	–	–	–	As above	440 0.50%	1–60	–	–	(ii)
–	–	–	–	–	–	As above	605 0.50%	1–60	–	–	(iii)
											n
–	–	–	–	–	–	As above	506 0.25%	1–60	–	–	(i)
–	–	–	–	–	–	As above	748 0.25%	1–60	–	–	(ii)
											p
–	–	–	–	–	–	As above	517 0.25%	1–60	–	–	(i)
–	–	–	–	–	–	As above	781 0.25%	1–60	–	–	(ii)
–	–	–	–	–	–	As above	847 0.25%	1–60	–	–	q
											r
–	–	–	–	–	–	As above	880 0.25%	1–60	–	–	(i)
–	–	–	–	–	–	As above	2,365 0.25%	1–60	–	–	(ii)

6.B Drainage — *continued*

No	Description		Capital	Cleaning	Cyclical	Repairs	Total
1	PIPEWORK — *continued*	10m					
s	Cast iron pipes and fittings; lead socketted joints; concrete bed and haunching; excavation						
	(i) 100 nominal; 750 deep		578.23	–	–	0.41	578.64
	(ii) 150 nominal; 750 deep		900.76	–	–	0.62	901.38
	(iii) 225 nominal; 1000 deep		2,422.86	–	–	1.67	2,424.53
2	ACCESSORIES	Nr					
a	Gulley, UPVC, trapped, 100 outlet, grating; concrete surround		30.67	10.47	–	0.09	41.23
b	Gulley, clayware, trapped, 100 outlet, grating; concrete surround		32.26	10.47	–	0.09	42.82
c	Gulley, clayware, trapped, reversible, 100 outlet, back inlet, hopper, grating; concrete surround		52.18	10.47	–	0.15	62.80
d	Gulley, cast iron, trapped, reversible, 100 outlet, back inlet, hopper, sealing plate; concrete surround		170.91	10.47	–	0.47	181.85
e	Yard gulley, clayware, trapped, 300 × 600, 100 outlet; concrete surround; medium cast iron grating; brick kerb		209.55	57.39	–	0.59	267.53
f	Yard gulley, cast iron trapped, 300 × 610, 100 outlet; concrete surround; medium cast iron grating; brick kerb		493.36	57.39	–	1.36	552.11
g	Road gulley, concrete, trapped, 450 × 750, 150 outlet; heavy duty grating; concrete surround; brick kerb		248.09	59.63	–	0.71	308.43
3	MANHOLES AND THE LIKE	Nr					
a	Polypropylene inspection chamber; 5 inlets; steel cover and frame; concrete surround; excavation; 475 diameter; 585 deep		105.83	–	–	0.30	106.13
b	UPVC shallow access chamber; 3 inlets; light duty cover and frame; concrete surround; excavation; 450 diameter; 600 deep		63.23	–	–	0.18	63.41
c	Vitrified clay inspection chamber; 4 inlets; alloy cover and frame; concrete surround; excavation; 450 diameter; 600 deep		88.99	–	–	0.27	89.26

Cleaning			Cyclical			Repairs					
Annual Cost	Frequency No	Yr	Work	Cost	Cycle in Years	Cause/ Action	Cost (£) and Risk	Occurs in Years	kg/m²	R value	No
											1
											s
–	–	–	–	–	–	Obstructions, subsidence, crushing, cracking, root entry	605 0.25%	1–60	–	–	(i)
–	–	–	–	–	–	As above	924 0.25%	1–60	–	–	(ii)
–	–	–	–	–	–	As above	2,475 0.25%	1–60	–	–	(iii)
3.69	1	5	–	–	–	Replace grating	33 1%	1–60	–	–	a
3.69	1	5	–	–	–	As above	33 1%	1–60	–	–	b
3.69	1	5	–	–	–	As above	55 1%	1–60	–	–	c
3.69	1	5	–	–	–	Replace sealing plate	176 1%	1–60	–	–	d
20.23	1	5	–	–	–	Replace or raise grating	220 1%	1–60	–	–	e
20.23	1	5	–	–	–	As above	506 1%	1–60	–	–	f
3.69	1	1	–	–	–	As above	264 1%	1–60	–	–	g
											3
–	–	–	–	–	–	Replace or raise cover	110 1%	1–60	–	–	a
–	–	–	–	–	–	As above	66 1%	1–60	–	–	b
–	–	–	–	–	–	As above	99 1%	1–60	–	–	c

6.B Drainage — *continued*

No	Description	Capital	Cleaning	Cyclical	Repairs	Total
3	MANHOLES AND THE LIKE — *continued* Nr					
d	GRP inspection chamber; 4 inlets; single seal cover and frame; concrete surround; excavation; 475 diameter; 940 deep	187.41	–	–	0.52	187.93
e	Cast iron inspection chamber; 4 inlets; bolted cover; bedding in cement and sand; 100 × 100	172.77	–	–	0.47	173.24
f	Brick manhole, class B engineering; concrete in situ base 150; concrete reinforced top 150; 4 inlets; channels; benching; single seal cover and frame grade B, 600 × 450; step irons; excavation					
	(i) 460 × 610 × 600 deep	365.38	–	–	0.50	365.88
	(ii) 1050 × 750 × 1200 deep	740.76	–	–	1.01	741.77
	(iii) 1350 × 1050 × 2550 deep	1,660.24	–	–	2.25	1,622.49
g	Concrete manhole, precast side units and top, jointed cement and sand; concrete in situ base 150; concrete surround; 4 inlets; channels; benching; single seal cover and frame grade B, 600 × 450; step irons; excavation					
	(i) 610 × 460 × 600 deep	305.75	–	–	0.43	306.18
	(ii) 1030 × 690 × 1200 deep	500.39	–	–	0.68	501.07
h	Concrete manhole, precast concrete base, side units and top, jointed cement and sand; concrete surround; 4 inlets; single seal cover and frame grade B, 600 × 450; step irons; excavation					
	(i) 675 diameter; 600 deep	289.50	–	–	0.40	289.90
	(ii) 1050 diameter; 1200 deep	580.87	–	–	0.80	581.67
	(iii) 1350 diameter; 2250 deep	1,571.45	–	–	2.13	1,573.58

Cleaning			Cyclical			Repairs					
Annual Cost	Frequency No	Yr	Work	Cost	Cycle in Years	Cause/ Action	Cost (£) and Risk	Occurs in Years	kg/m²	R value	No
											3
–	–	–	–	–	–	Replace or raise cover	193 1%	1–60	–	–	d
–	–	–	–	–	–	Replace	176 1%	1–60	–	–	e
											f
–	–	–	–	–	–	Replace or raise cover	374 0.50%	1–60	–	–	(i)
–	–	–	–	–	–	As above	748 0.50%	1–60	–	–	(ii)
–	–	–	–	–	–	As above	1,672 0.50%	1–60	–	–	(iii)
											g
–	–	–	–	–	–	As above	319 0.50%	1–60	–	–	(i)
–	–	–	–	–	–	As above	506 0.50%	1–60	–	–	(ii)
											h
–	–	–	–	–	–	As above	297 0.50%	1–60	–	–	(i)
–	–	–	–	–	–	As above	594 0.50%	1–60	–	–	(ii)
–	–	–	–	–	–	As above	1,584 0.50%	1–60	–	–	(iii)

6.C External services

No	Description	Capital	Cleaning	Cyclical	Repairs	Total
1	WATER MAINS 10m					
	10m Pipe and fittings; valve; stopcock pit and cover					
a	Polyethylene, B.S. 6572 blue; trench 750 deep; backfill; 51 diameter	342.25	–	2.07	0.49	344.81
b	Galvanised steel, BS 1387; anit-corrosive wrapping; trench 750 deep; backfill; 50 diameter	358.41	–	2.07	0.52	361.00
c	Copper, BS 2871, Table Y, plastic coating; trench 750 deep; backfill; 54 diameter	495.63	–	2.07	0.71	498.41
d	Cast iron, BS 4772, ductile spigot and socket joints, Class K9; trench 750 deep; backfill; 80 diameter	603.08	–	2.07	0.89	606.04
e	Cast iron, BS 4772, flanged joints, Class K12; trench 750 deep; backfill; 80 diameter	645.04	–	2.07	0.96	648.07

Cleaning			Cyclical			Repairs					
Annual Cost	Frequency No	Yr	Work	Cost	Cycle in Years	Cause/ Action	Cost (£) and Risk	Occurs in Years	kg/m²	R value	No
											1
–	–	–	Rebed pit cover	11.88	30	Damage	363 0.50%	1–60	–	–	a
–	–	–	As above	11.88	30	As above	385 0.50%	1–60	–	–	b
–	–	–	As above	11.88	30	As above	528 0.50%	1–60	–	–	c
–	–	–	As above	11.88	30	As above	660 0.50%	1–60	–	–	d
–	–	–	As above	11.88	30	As above	715 0.50%	1–60	–	–	e

6.C.8 Builder's work in connection with external services

No	Description	Capital	Cleaning	Cyclical	Repairs	Total
1	CABLE DUCTS 10m					
	10m duct including bends and ends where required.					
a	UPVC pipe and fittings, BS 6640, ringseal joints; trench 750 deep; backfill					
	(i) 100 nominal	143.67	1.63	–	0.22	145.52
	(ii) 150 nominal	190.25	1.63	–	0.28	192.16
b	Vitrified clay pipe and fittings, BS 65, extra strength, sleeve joints; trench 750 deep; backfill					
	(i) 100 nominal	136.26	1.63	–	0.22	138.11
	(ii) 150 nominal	174.24	1.63	–	0.27	176.14
c	Vitrified clay pipe and fittings, BS 65, extra strength, sleeve joints; trench 500 deep; in situ concrete surround					
	(i) 100 nominal	218.49	1.63	–	0.33	220.45
	(ii) 150 nominal	286.19	1.63	–	0.43	288.25
d	Precast concrete pipe and fittings, BS 5911, Class M, flexible joints; trench 750 deep; backfill; 300 nominal	344.56	1.63	–	0.50	346.69
e	Precast reinforced concrete duct and cover; interlocking joints; trench 250 deep; in situ concrete bed and haunch; 100 × 100 internally	194.58	1.63	26.79	0.30	223.30
f	In situ reinforced concrete duct and cover, trench 750 deep, 150 base, 100 walls, 100 precast cover, jointed; 600 × 500 internally	1,893.93	–	–	2.67	1,896.60

Cleaning			Cyclical			Repairs					
Annual Cost	*Frequency No*	*Yr*	*Work*	*Cost*	*Cycle in Years*	*Cause/ Action*	*Cost (£) and Risk*	*Occurs in Years*	*kg/m²*	*R value*	*No*
											1
											a
Rodding out when rewiring 1.36	1	10	–	–	–	Damage	165 0.50%	1–60	–	–	(i)
As above 1.36	1	10	–	–	–	As above	209 0.50%	1–60	–	–	(ii)
											b
As above 1.36	1	10	–	–	–	As above	165 0.50%	1–60	–	–	(i)
As above 1.36	1	10	–	–	–	As above	198 0.50%	1–60	–	–	(ii)
											c
As above 1.36	1	10	–	–	–	As above	242 0.50%	1–60	–	–	(i)
As above 1.36	1	10	–	–	–	As above	319 0.50%	1–60	–	–	(ii)
As above 1.36	1	10	–	–	–	As above	374 0.50%	1–60	–	–	d
As above 1.36	1	10	Reset covers	22.40	10	As above	220 0.50%	1–60	–	–	e
–	–	–	–	–	–	As above	1,980 0.50%	1–60	–	–	f

MEMORANDA

Thermal Resistance

Surface resistances to be added to the resistances of the materials calculated from the Tables.

OUTSIDE SURFACE

ELEMENT	EMISSIVITY OF SURFACE	Surface Resistance (m^2k/w)		
		Sheltered	Normal	Severe
Walls	High Low	0.08 0.11	0.06* 0.07	0.03 0.03
Roofs	High Low	0.07 0.09	0.04 0.05	0.02 0.02

INSIDE SURFACE

ELEMENT	HEAT FLOW	Surface Resistance (m^2k/w)	
		High emissivity factor	Low emissivity factor
Walls	Horizontal	0.12*	0.30
Ceilings or roofs, flat or pitched; floors	Upward	0.10	0.22
Ceilings and floors	Downward	0.14	0.55

* Used in the windows and doors calculations in the Tables.

For further information the user is recommended to consult CIBSE GUIDE A3 — Thermal Properties of Building Structures — 1980, reprinted 1985 and published by The Chartered Institution of Building Services Engineers.

Discount Factors

Present Value of future payments of £1.00 at typical discounts for building of 60 years life. Assumed that renewals would not be made in the final year.

	3%	6%	10%
Every year to 60	27.676	16.161	9.967
Every second year to 58	13.464	7.815	4.743
Every third year to 57	8.784	5.046	3.008
Every fourth year to 56	6.445	3.664	2.144
Every fifth year to 55	5.043	2.837	1.629
Every fifth year to 55 commencing at year 15	3.436	1.531	0.623
Every sixth year to 54	4.109	2.287	1.289
Every sixth year commencing at year 3	4.675	2.760	1.719
Every seventh year to 56	3.519	1.910	1.049
Every tenth year to 50	2.244	1.196	0.622
Every twelfth year to 48	1.780	0.928	0.463
Every fifteenth year to 45	1.318	0.664	0.310
Every eighteenth year to 54	1.135	0.516	0.218
Every twentieth year to 40	0.860	0.409	0.171
Every twentyfifth year to 50	0.706	0.287	0.101
Thirtieth year	0.412	0.174	0.057
Fourtieth year	0.307	0.097	0.022
Forty-fourth year	0.272	0.077	0.015
Fiftieth year	0.228	0.054	0.009

LOCATION STUDY MAP

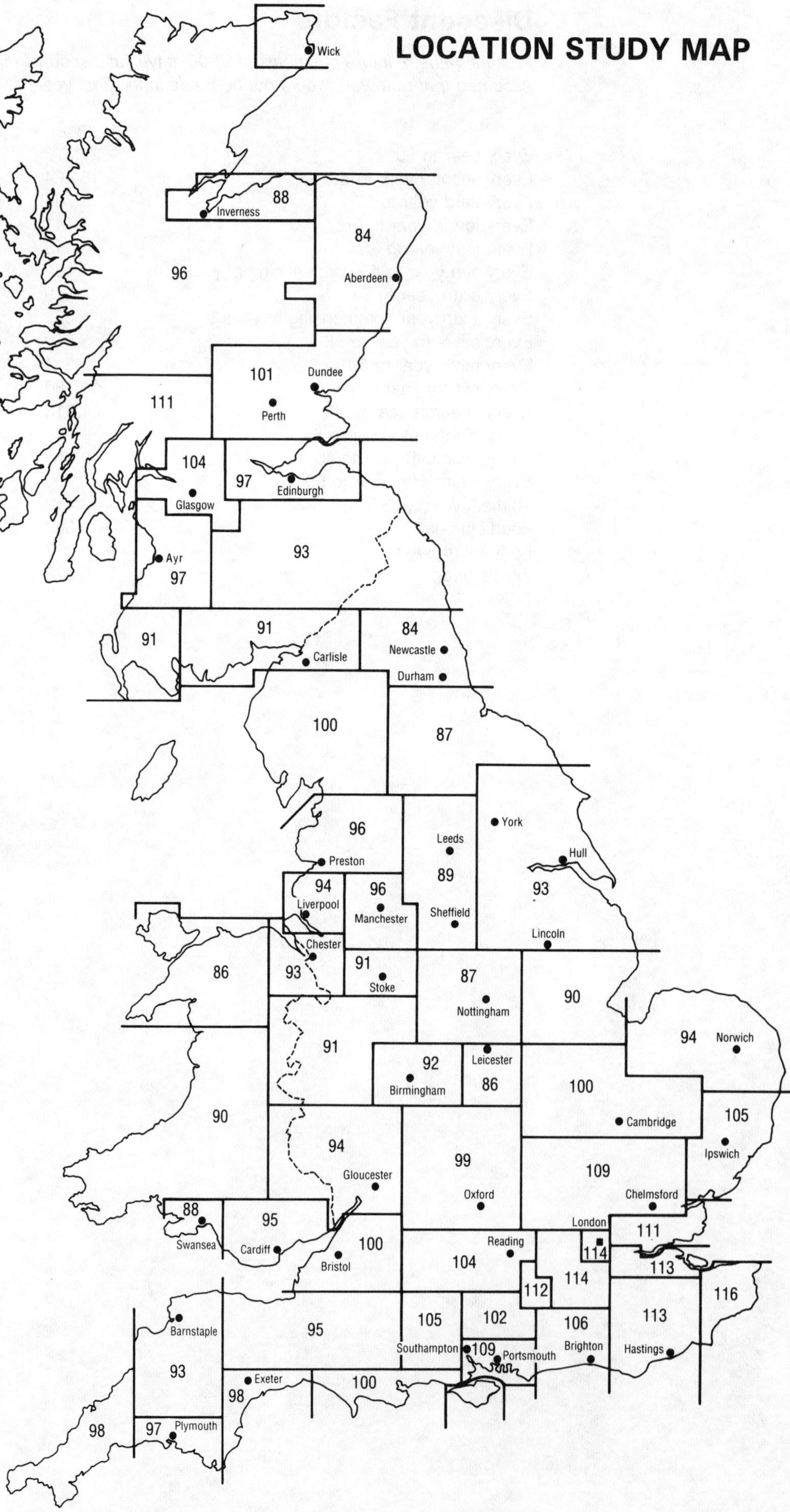

Location Factors

Title of Area	Location Factor	Trimmed (Interquartile) Range
Shetlands	—	—
Orkney	—	—
Wick/Thurso	—	—
Moray Firth	88	76– 90
Aberdeen	84	72– 94
Rest of N Scotland	96	92–119
Argyll	111	91–129
Perth/Dundee	101	77–112
Glasgow	104	93–114
Edinburgh	97	84–113
Ayr	97	73–121
Borders	93	86–119
Wigtown/Stranraer	91	84–103
Dumfries/Carlisle	91	84–103
Tyne & Wear	84	77– 96
Cumbria	100	90–108
Durham/Cleveland	87	74–100
Lancashire	96	82–109
Merseyside	94	80–100
Gtr Manchester	96	86–107
Leeds/Sheffield	89	75–100
York/Humberside	93	81–106
Gwynedd	86	82– 93
Cheshire	93	78–116
Stoke-on-Trent	91	73–109
Derby/Nottingham	87	76–103
Lincolnshire	90	85– 99
Mid and West Wales	90	87–100
Shropshire	91	81–105
Birmingham	92	80–107
Leicester	86	79– 98
Cambridge	100	86–114
Norfolk	94	82–105
Gloucester/Worcester	94	78–101
Oxford	99	89–115
Beds & Hertfordshire	109	95–121
Colchester/Suffolk	105	90–121
Swansea	88	82– 94
Glamorgan	95	83–111
Bristol/Bath	100	84–114
Swindon/Reading	104	94–119
Aldershot	112	94–121
South and West London	114	103–129
London Central/Postal	114	101–134
North Thames	111	96–131
South Thames	113	98–134
Cornwall	98	88–105
North Devon	93	89– 98
Plymouth	97	89–104
Taunton/Yeovil	95	79–105
Exeter/Torbay	98	84–106
Dorset	100	87–117
Salisbury	105	89–118
Hampshire	102	93–115
So'ton/Portsmouth	109	90–124
Isle of Wight	—	—
West Sussex	106	99–114
East Sussex/Weald	113	97–126
East Kent	116	107–126

Abbreviations

dp	=	deep
GFA	=	gross floor area
hy	=	high yield (reinforcement)
K	=	thousand
N	=	N/mm2
RSC	=	rolled steel channel (structural steelwork)
sw	=	softwood
t	=	tonne
t & g	=	tongued and grooved
thk	=	thick
UB	=	universal beam section (structural steelwork)

Printed in the United Kingdom for HMSO
Dd.291981 4/91 C21 3385/4 5673 118849